Journey
to the Tar Sands

Jeh Custer, Gregory John Ellis, Aftab Erfan, Jacqueline Gamble,
Kealan Gell, Jodie Martinson, Tim Murphy, Dylan Sparks, Kalin Stacey,
Lindsay Telfer, Katherine Trajan, Lori Theresa Waller

Editorial coordination by Tim Murphy

James Lorimer & Company Ltd., Publishers
Toronto

To those we met along the way.
Thank you for your hospitality &
thank you for sharing your stories.

This book is printed on Neo Gloss paper, which is FSC (CoC) Mixed Sources
accredited. The production of Neo uses fibres from sources approved by FSC, PEFC
or CSA and operates under the framework of ISO14001 environmental standards. An
environmentally friendly three step waste water treatment process is followed, ensuring
that minimal impact is made on local waterways. All pulp used is elemental chlorine free.

Portions of "Disposable Workers and Labour Rights" by Tim Murphy and "Passing Out in
Upgrader Alley" by Lori Waller first appeared in slightly different form in The Dominion Paper's
feature Tar Sands issue, available online at http://www.dominionpaper.ca/print/tar_sands_issue_48.

Photo Credits:
Photographs of Journey to the Tar Sands were supplied by the riders and coordinated by Jacqueline Gamble.
The map appearing page 4 was produced by Jeh Custer. The authors and publisher thank David Dodge and the
Canadian Parks and Wilderness Society for supplying the images of the tar sands appearing on pages 82–83; the
Pembina Institute for those appearing on pages 85 and 92; and Garth Lenz (www.garthlenz.com) for the image page 84.
Other images: 93, The Canadian Press/Edmonton Sun, Jordan Verlage; 86, gettyimages, Sarah Leen; 25, 61, istockphoto.

James Lorimer & Company Ltd., Publishers acknowledge the support of the Ontario Arts Council. We acknowledge the support of the
Government of Canada through the Book Publishing Industry Development Program (BPIDP) for our publishing activities. We acknowledge the
support of the Canada Council for the Arts for our publishing program. We acknowledge the support of the Government of Ontario through the
Ontario Media Development Corporation's Ontario Book Initiative.

 Canada Council   Conseil des Arts
for the Arts   du Canada

ONTARIO ARTS COUNCIL

Library and Archives Canada Cataloguing in Publication

Journey to the tar sands / edited by Tim Murphy.

ISBN 978-1-55277-039-9

1. Oil sands—Environmental aspects—Alberta—Fort McMurray Region.
2. Oil sands—Alberta—Fort McMurray Region.   3. Sierra Youth Coalition.
4. Bicycle touring—Alberta.   5. Alberta—Description and travel.   I. Murphy, Tim

TD195.P4J682 2008   333.8′232097123   C2008-903582-8

James Lorimer & Company Ltd., Publishers
317 Adelaide Street West, Suite 1002
Toronto, ON M5V 1P9
www.lorimer.ca

Printed in China

Contents

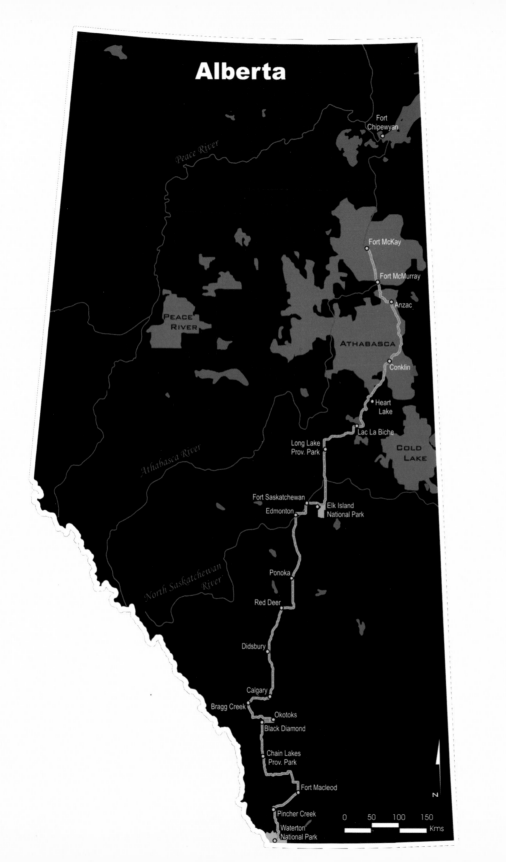

Foreword

The climate change issue has a lot of "canaries in coal mines" — the loss of Arctic ice, the rising seas threatening to inundate South Pacific nations, the cracks in Antarctic ice shelves. Photographs of glaciers in retreat, of advancing deserts, of endangered polar bears are the iconic images of the coming climate crisis. We see pictures of the impacts.

To find a picture of the cause — a big and ugly iconic image of the worst of humanity's addiction to fossil fuels — the violent scraping away of the boreal forest and the creation of cavernous open-pit mining in the Athabasca tar sands is the most likely candidate.

Tim Murphy writes of it in this volume as the "orifice" and as "the belly of the beast." It is, in any case, the cause of the worsening record of Canada's rising emissions, as we march steadfastly away from the commitments made under international law to reduce emissions.

The threat of an abrupt and devastating shift in global climate is increasing. No place on Earth will be spared the impacts of a destabilized climate. Alberta faces a serious water crisis as the glaciers retreat and evaporation increases. The Athabasca River is already experiencing unprecedented low flow levels. More persistent droughts loom on the horizon in the province where provincial governments have denied the threat of climate change, while designing "make-believe" strategies for intensity improvements.

The international scientific community is increasingly clear. We are running out of time if we are to avoid irreversible and ever-worsening climate disaster globally. In fact we have only until 2015 to ensure global emissions of greenhouse gases peak and begin to fall everywhere.

Canadians cannot pretend the damage to the global atmospheric commons caused by the tar sands is acceptable. We cannot turn the other way. Raising awareness and understanding the threat are key. It is to this end that the wonderful young people of Sierra Youth Coalition undertook their tour of the tar sands.

They take us on the tour of ranchlands compromised by oil development, to the promise of the future in wind mills and renewable energy, with side trips into global trade deals and the Security and Prosperity Partnership. The aching limbs and hunger along the way all add up to the commitment of youth to change course.

We, of the generation that most contributed to the global disaster of climate change, cannot help but be inspired by the activism of youth. As we read the chronicles of the journey to the source of a threat to our collective future, we must re-commit to match the energy and dedication of these young people. Not all of us can bike the Rockies to take on the tar sands, but we can commit to the collective effort that will be required to avoid runaway global warming. We do not have much time to shift to a low-carbon future. But we do have some time. We are running out of excuses. Pick up this book to start your journey.

— *Elizabeth May & Maude Barlow*

An Idea Takes Shape

A journey to the orifice, we jokingly called it. That was it. We would travel to this godforsaken, larger-than-life place we'd only heard of — the Alberta tar sands, a vast northern area the size of Florida and rich in a sticky substance called bitumen, used to create synthetic oil. And arguably the single largest obstacle to Canada's meeting its targets under the Kyoto Protocol.

In 2007, more and more Canadians were being galvanized into their own personal "green revolution" by their growing concern over climate change and their desire to safeguard the planet for future generations. In particular, the youth environmental movement had identified the climate crisis as the defining issue of its generation.

Yet at the same time, it seemed our government in Ottawa was moving in the other direction, edging closer to failing its real, multilateral environmental commitments. And far away from central Canada, in the wild west of "Boomtown," money was oozing from the ground like black gold. The Alberta tar sands were the fastest-growing source of Canada's greenhouse gas emissions and were expected to overtake coal as its leading source of emissions.

The Sierra Youth Coalition (SYC), one of Canada's leading youth environmental groups, decided that its work on the climate change file would not be complete until it addressed the unabated development occurring in Alberta's oil patch. Acknowledging the complexity of energy production and the difficulty of comprehending the scale and implications of rushed industrial development, we decided to go and see it for ourselves. We would go so that we could better understand the challenges Canada faced, and so that we could help other Canadians like us understand. We weren't interested in merely parroting the message of others — we wanted to make a unique contribution to the public debate. After consulting with activists from Alberta, we made our decision.

SYC has a history of activist caravans. In 2001, the Climate Change Caravan cycled from Tofino, B.C., to Halifax, N.S., in the hopes of inspiring Canadians to reduce their personal greenhouse gas emissions. In 2003, youth from across Canada set out from Vancouver, B.C., bound for the 5th Ministerial Conference of the World Trade Organization in Cancun, Mexico, 3,000 kilometres away, to raise awareness about the social and ecological impacts of trade liberalization, especially in the area of food production. It was time to get the bikes on the road again: we would cycle to the tar sands looking for the facts and the stories of those most affected by the industry.

As a former SYC employee, an outgoing member of its Executive Committee, and a long-time member and participant in campaigns such as the Climate Change Caravan, Tim was well positioned to take the lead in organizing. In March of 2007, after many months of deliberation and debate, we took the plunge. We sent out a call for participants and received well over 50 responses. In the end, only 19 would join us on the journey. Our decision to make the bike trip application process as simple and open-ended as possible meant that people were joining our group mere weeks before the scheduled date of departure, while others dropped out as the trip drew near. Our attempt at organizing working groups to tackle various aspects of the preparation for the trip fell apart early on, as most riders were busy with summer jobs or off on summer adventures of their own. As a result, the organizing fell to Tim and a few other committed individuals. Luckily, Tim had the support of the Sierra Club's Prairie Chapter, whose director, Lindsay Telfer, a past director of SYC, was busy fighting to put a stop to the unchecked exploitation of Alberta's tar sands. She would help us secure funds to hire Shawn, one of the bikers, to help plan the logistics of the trip.

To raise money for the trip, we gave each rider the goal of raising $500: some exceeded this goal, others didn't even come close. In the end, we succeeded in raising a modest, yet respectable $8,000, mostly through donations from family and friends. It wasn't much, but coupled with the generosity of the individuals we met along the way, it would see us through.

The members of our group came from across the country, from Vancouver to Halifax, and from as far away as California and Massachusetts. Together, we would cycle from Waterton-Glacier National Park, on the U.S.—Alberta border, due north to the Native community of Fort McKay, just north of Fort McMurray. We would visit cities such as Calgary and Edmonton, rural communities, working towns and First Nation reserves. We mapped our route by connecting the dots representing communities relevant to the subject at hand (Pincher Creek for its renewable energy projects; Turner Valley because it is the site of Alberta's first oil boom; Okotoks as a model for sustainable urban development; Fort Saskatchewan as an example of a community affected by the industry around oil production; and so on). We then searched out contacts in each community who would help with food, accommodations, media and setting up interviews or meetings with the locals. They would be invaluable to our efforts.

Although as organizers we'd attempted to sell the trip as a storytelling adventure and fact-finding mission, our group was by no means unanimous on our purpose for being there. At times there seemed to be as many interpretations of our mission as there were people on the actual trip. Some of us had come to learn a bit more about the tar sands, some of us were drawn by the challenge of long-distance cycling, some of us found the trip's communal aspects most

Discussing our media strategy

appealing, some of us liked its non-preachy and open-minded approach, and still others hoped it would spark a broader movement, bring down the man and contribute to stopping tar sands development once and for all. Similarly, some of us were well versed in tar sands vernacular, while others could barely find the area on a map; some of us were formidable athletes with extensive cycling experience, while others were wheezing after the first 10 minutes. Still, while we differed on many fronts, we all knew we'd be talking to countless people along the way, and that our job was to get their stories out in whatever way we could. If we agreed on one thing, it was that there was more to the tar sands story than what we were being fed by the mainstream media.

As observers, our job was to seek the truth. The hardest part of that job was letting go of preconceived ideas of what that truth might be, and being open to others' points of view. Our views were coloured by our own experience and although we might profess objectivity, our bias remained — like a pair of tinted glasses. To get

beyond it, we had to be willing to change our glasses and view a situation from another perspective. From the outset, our goal to remain objective would be pitted against our penchant for activism and our various opinions on the situation at hand. To the very end, we would struggle to find a common message, often contradicting each other in our dealings with the media and the public at large. This seems to reflect people's diverse responses to environmental problems in general. Engineers, ecologists, geographers, economists, politicians, pipefitters and activists each have a language of their own that shapes their view.

We would aim to start a conversation between the many "stakeholders" involved: farmers and ranchers, executives and oil workers, bureaucrats and politicians, men and women, First Nations, Albertans, new Canadians, old Canadians, Americans and other citizens of the world. We wanted to encourage debate and discussion, rather than put people on the defensive and shut down real communication. We wanted to hear peoples' stories and ideas and share our own. For far too long, it had been almost taboo to question the merits of oil development in Alberta, and we were providing a soapbox for other points of view. So it should be no surprise that most of our stories come from those experiencing the downside of tar sands development. Those who preach the economic benefits of tar sands development are quoted weekly in the business sections of national newspapers. They don't need our help.

This is the story of our journey to the tar sands and what we learned along the way.

Week One: Waterton-Glacier to Bragg Creek

Greyhound lost Kalin's bike!

Kalin and Lindsay help Shawn put his bike together

August 15
Meeting Up in Waterton-Glacier

Exhausted, I stepped off the Greyhound in Pincher Creek, Alberta, after nearly three days travelling westward from Montreal; I was as close as I could get to our arranged meeting place, Waterton-Glacier Park. It was well after midnight and the town was fast asleep. With my bags strewn about the dark parking lot of the local flower shop/bus station, I sat under a street light and pieced together my red hand-me-down bike and attached the trailer ready to haul communal items such as camping stoves, food, bike tools, pots and pans — everything including the kitchen sink (in our case a small plastic dish pan, some biodegradable dish soap and a few sponges). The bike would also carry two small saddlebags filled with personal gear and my laptop. Bike intact, I rolled around the block to the neighbouring school ground, nestled my tent up against the building, looked up at the starry sky and breathed a sigh of relief: I had arrived. I slipped into my sleeping bag and closed my eyes, only to be awoken much too soon by the morning sun pounding down on the tent, making it unbearably hot. The building was not a school, but a community centre. I emerged from the tent and peered down through its large picture window at a woman swimming laps in the pool.

On this very day, former premier of Alberta Peter Lougheed made national news by warning that a constitutional debate over the Alberta oil sands, pitting the province's right to develop its mineral resources against the federal government's responsibility to curb greenhouse gas emissions, was in the offing. The headlines were calling for a national debate on the oil sands and we were on the ground, getting it started. The adventure had begun.

As I zipped back by the bus station on my way out of town, I spotted Kalin, whom I'd known would be travelling more or less the same route as me. From Toronto, Kalin is our iPod-clad city-boy of the group; in days ahead he would lead us

Gathering at the entrance to Waterton

in meditations and bound joyfully over bales of hay when the spirit moved him, but at the moment he was sitting dejectedly on his luggage, no bike in sight. With his numerous suitcases he looked more like a stranded passenger in an airport than someone about to embark on a three-week bike trip. Greyhound had lost his bike somewhere on the journey; and he'd misplaced his wallet. He was understandably distressed. Riding with "To the Tar Sands" was Kalin's first real foray into the world of activism. His fervent dislike of Greyhound would quickly be overtaken by a vehement reaction against the faceless corporations plundering Alberta, as he became an empassioned activist.

I sat with Kalin as he made to calls to riders travelling to the meet-up by car and arranged a lift, imagining the others, on buses, trains, cars and planes, having similar adventures as they converged from all parts of Canada to form the original group of nine that would cycle more than 1,500 kilometres due north from the Alberta–Montana border to the Athabasca Tar Sands. With Kalin's arrangements made, I got on my bike and set out to cycle the extra 50 kilometres from Pincher Creek to meet my group in

Waterton-Glacier, riding along the Cowboy Trail through golden fields of barley set against a rugged mountain backdrop, imagining tumble-weed and watering holes, and occasionally letting loose a "yeeee haaaa." I was overjoyed.

The last downhill into Waterton-Glacier is a long one. It's the kind of hill that both thrills and dismays, for while you revel in the speed of the descent, you wince at the thought of having to climb your way back out. I was the last to arrive. The group was there to greet me at the bottom of the hill. I saw Aftab, then Kealan, Lindsay, Shawn, Jodie, Jackie and even bikeless Kalin, who'd skipped ahead of me after hitching a ride with Jodie's parents. Everyone was there except Dylan, who was said to be somewhere in the park.

"We picked up one more rider," I heard someone say. That rider was Greg, a California student, originally from Wyoming, whom Kealan had met on the train ride up from Michigan and promptly invited along. Be it a tribute to Kealan's charm or Greg's whimsy, the boy was immediately sold on the idea and in a blind leap of faith decided to forego the bike trip through the upper United States he'd planned and head north to the tar sands instead.

A beautiful thing about having Greg join us was that he, more than anyone, was coming into this with an open mind, a fresh perspective. I tried not to laugh when I heard Greg speak of heading to "Fort McMurphy," and tried not to show my glee at having an American on board.

Greg is quicksilver, quiet and introspective one moment, exuberantly, comically dancing the next, flailing his arms about like an octopus. During the trip he will summon us to group meetings with a tin whistle he picked up along the way. Greg thinks in riddles, and sees our mission as one akin to healing: we are here to do no harm, but to spread the word of a sickness spreading among us — the devastation occurring in the New Canadian West — in the hopes that we can stem it before it becomes epidemic.

Aftab is a beautiful person, radiant. Originally from Iran, she has that perfect blended accent that says, "I am from here, but I was not born here." The kind of accent the CBC loves, or so they told her once. And love her they would. Cycling next to Aftab, you were just as likely to catch her fumbling for her cell phone as reaching for her water bottle. She became notorious for fielding half-hour media calls from random ditches on the sides of highways.

By default, those of us carrying cell phones and laptops became spokespeople for the group. We responded to requests from media as best we could, charging our phones in public washrooms and sending emails after hours from the curb in front of small-town libraries, hitching a ride on their wireless Internet.

We found Dylan on the road as we made our way to the campground. Charming and good-looking, Dylan would become the trip's "poster boy." A Vancouverite, he was just 18, one of the youngest in our group, which may explain his voracious appetite. An experienced outdoorsman, he was supremely comfortable out of doors, dropping almost instantly to sleep when he felt tired, like a wild creature, under the starry sky, in an open field. We would discover that he howled with the wolves and at the moon. One day, Aftab and I spotted him up ahead sprawled across the pavement. Thinking he'd taken a fall, we raced to his rescue. We were both relieved and annoyed when we found him intent on taking a photograph of a bee on the shoulder of the road. Dylan documents every occasion, in his journal or through photographs — many of which appear in this book.

— Tim

Tonight's Digs: Camping in Waterton Lake National Park.

Mulling over the route

This is how Aftab explained Deep Democracy to us — a philosophy that would ultimately come to guide our interactions on the trip. "In every group, there are roles to be filled. All of us come to embody a certain role, but the roles exist independent of one's presence in the group. Were a group member to leave, someone else would step up to fill that role. And so there is a role for the joker, a role for holding information, for planning logistics, for being negative, for cooking great food, for feeling tired. We must acknowledge and honour these roles."

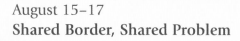
Our point of departure

Signs like this dotted southern Alberta ranchland

August 15–17
Shared Border, Shared Problem

Our point of departure was carefully chosen. Waterton-Glacier is the world's first international peace park. Straddling the Canadian–American border, we saw the park as a symbol of the complicity between the two countries in the rush to extract tar sands oil, as well as the co-operation that will be required to end our joint dependence on fossil fuels.

Currently, 70 percent of the crude oil extracted from the Alberta tar sands heads south of the border. This percentage is expected to increase with the completion of projects such as the proposed Keystone Pipeline, which would traverse the province and cross the midwestern United States carrying upwards of 400,000 barrels of bitumen and unrefined crude oil a day from the Alberta oil fields to refineries in Illinois.

From our campsite, we could see small patches of snow up on the mountaintop, snow that had somehow managed to survive the hot summer sun. Most of the remaining glaciers are now found on the Montana side of the park but they are melting at an alarming rate. In 1850, the park had 150 glaciers. Today, because of global warming, there are only 27 left, and the best predictions suggest that all the glaciers in the park will be gone by the year 2030.

Mountain view, Waterton

Waterton-Glacier is an awe-inspiring place. It stood in stark contrast to what we would encounter when we finally reached the tar sands. Until recently, the northern boomtown of Fort McMurray was as pristine a place as Waterton Park, but with the current tar sands expansion, air quality in the city of 70,000 is said to be as bad as the large metropolises of Edmonton and Calgary.

There are oil and gas developments speckled throughout the region surrounding the park, including a large Shell sour gas plant nearby. A sign at the lookout near the entry to the park describes how they have worked to achieve harmony between our society's need for fossil fuels and our need to protect the majestic natural areas that we all love to visit, the health of which we ultimately depend on for our own. Reading that sign led me to wonder whether such harmony will ever come to the tar sands.

The peace park was born out of the Great Depression, shortly after the First World War. Ironically, in a time of economic crisis and social hardship, our two countries were able to recognize the importance of joining the two parks for the future benefit of the species that inhabit them (wildlife is generally ignorant of human borders). Today, in a time of great prosperity, we are developing the tar sands as though there will be no tomorrow, forgetting that when the oil sands have been all dug up, and our rivers drained dry and our bank accounts emptied, the mess will still remain for future generations to clean up.

The challenges of sticking to a schedule that had been drafted in front of a computer screen were apparent from that very first day. We were supposed to spend that night on the Diamond Willow Beef Ranch near Twin Butte, halfway between Waterton National Park and Pincher Creek, and interview the owner, Larry Firth. Meanwhile, we had three interviews scheduled in

We spent our first two days in Waterton Park mulling over the route, establishing group process, sorting through equipment, taking inventory of our food supplies, discussing media strategies and learning as much as we could about the tar sands, while slowly getting to know each other. Ready or not, we departed from the park on August 17 as planned.

— Tim

Pincher Creek late that afternoon. There was no way we could make it into town for the interviews and cycle the 25 kilometres back to the ranch before dark. So we split up.

Jodie, our filmmaker, went off on her own to get some landscape shots for the film. She had joined the trip with the express purpose of documenting the experience. Jodie is from Calgary, and the story of the tar sands was one she had been itching to tell for some time — this was her chance.

There could be only one thing better than having an American on board and that was having a dissenting Calgarian, ready to question and challenge her fellow Albertans on their ideas of wealth and progress.

Aftab cycled ahead to find cell reception so that she could catch a conference call for her job as an urban planner back home in Halifax. Lindsay drove home to Edmonton and her job as director of the Sierra Club of Canada's Prairie Chapter, but not before helping Kalin retrieve his bike. I said goodbye, grateful for the support she'd given us as we prepared to leave, knowing that her briefing on tar sands activity would serve us well.

Meanwhile, Kealan, a B.C. native whom I'd first met back in Montreal when we were both students at McGill, had the honour of pulling one of

Jodie, our filmmaker

two loaded bike trailers on that first gruelling day. I would pull the other. Far ahead, he and Greg turned down a country road to the ranch, where they met with Larry and his wife, Jan; they would catch up with us later in Pincher Creek.

Soon thereafter, Shawn, Jackie, Dylan and I were the last to set off for Pincher Creek. Although our numbers would swell to 19 by the trip's end, this was our initial team. Small and a little scattered, but energized and raring to go.

Shawn is an unforgettable character. He is small, shy and awkward, yet stubborn and determined. He is a vegan and decidedly remained so despite the omnipresence of beef here in cattle country. One of the main organizers of the trip, he moved to Edmonton from Toronto to work for the Sierra Youth Coalition for minimum wage in a province where flipping burgers can earn you almost twice that. He spent the summer building our website and working out the logistics for the trip. Unfortunately, training for the bike trip was the one thing he'd apparently overlooked.

Jackie, Dylan and I kept pace with Shawn. By 3 p.m., on that scorcher of a first day, we were still only halfway to Pincher Creek and were moving at an average speed of seven kilometres an hour. We'd strapped Shawn's panniers onto our own loaded bikes and we walked by his side, cheering him up the hills. The slower speed allowed Jackie, Dylan and me to get acquainted. Jackie, I discovered, was from Toronto and had just recently returned from an internship in the Sudan. At 30, she was the oldest member of our group, but her spirit was as young as any. We would need her spirit soon, as a CTV news van from Lethbridge pulled up beside us to film the "caravan" of cyclists headed to the tar sands.

Shawn crests a hill on that scorcher of a first day

Jackie and I spoke to the camera, inaugurating our media spiels, while Dylan stood snacking in the background, with his shirt off and tucked down his shorts in a ridiculous attempt to cushion his saddle sores. Shawn was a small speck in the distance, slowly meandering his way up the long steady hills of that first day's ride. Later, when he finally turned to us and said he'd had enough, we immediately stuck our thumbs out and caught him a ride into town with the first passing truck. All bets were off as to whether Shawn would succeed in making it to Fort McKay. There's the turtle, there's the hare, and then there's Shawn.

— Tim

Our first news coverage!

Saddlebag inspector at
Gardners' ranch

August 17
Pedalling Through Ranch Country

As the others headed to Pincher Creek, Kealan and I veered off the pavement and onto a washboard gravel road that rolled over the hilly country. Not yet one day into the trip, we were about to have our first encounter with the Albertan oil and gas industry. One flat tire later, covered in road dust, we pulled into Larry and Jan's driveway.

Before we got down to business, the Firths insisted on treating us to a country-style lunch. We sat down to a spread of Tabor corn, garden fresh salad, potatoes and, the pièce de résistance, a slab of certified organic, grass-fed beef from one of their own herd. Now, I'm a vegetarian, but craving protein after the strain of biking and confronted with a steak in its most environmentally friendly and ethically sound form, not to mention cooked to perfection, my principles shrugged "Why not?" and let my stomach have its way.

After lunch, while we were still picking bits of steak from our teeth, Larry led us through knee-deep grass to the middle of one of his pastures. He pointed out encroaching developments — a survivalist neighbour's cabin on the western horizon, and in the windswept distance, an industrial complex for removing sulphur gas from drilled methane.

Indeed, Larry does not own as far as the eye can see, but even if he did, ownership offers only an illusion of security. Under Canadian law, a landowner's right to his land extends only five inches below the surface — the Crown (Alberta's provincial government in this case) retains rights to all resources below that level. Larry told us that the region has caught the fancy of all sorts of prospectors, seeking everything from gas and oil to diamonds to uranium. Alarmingly, the extraction rights for these resources, as well as easements for pipelines and power lines, can be auctioned off to the highest bidder without the owner's consent or even prior notification.

The seizure of land by a corporation begins long before shadowy landmen, the corporate agents or "mercenaries" (as the landowners call them), appear to negotiate terms of use and compensation with landowners. These agents are often less than forthcoming about the owner's rights, such as the right to choose the location of the alterations and to negotiate the value of leases. Landowners are easily and regularly taken advantage of, because they are generally unfamiliar with the formidable complexities of expropriation law, and unaware of the current market value of the leases, and there is no good source for advice. Landmen also prefer not to mention the precautions that can be taken by landowners, such as having their drinking-water well tested before drilling commences so that if contamination of the aquifer occurs they can prove that it is a result of the drilling. To further stymie landowners, Alberta law does not require an assessment to determine the pre-development state of the property before a lease is granted. Therefore it is often difficult to prove damages.

Larry insisted that the burden of proof for suspected contamination should be shifted to the corporations to better protect the landowners, since it is unrealistic and impractical to expect every landowner to find the time to become thoroughly educated in these complex issues. And his experience bears out his point of view. He gestured to an adjacent pasture, directing our attention to an encroachment on his own property — a collection of concrete pads, a rusting wellhead and a pipeline access-point marring the natural fescue grassland. Euphemistically dubbed "jewellery" by the energy corporations who own them, these decaying relics, already a familiar part of the landscape, may become permanent

The "jewellery" at Diamond Willow Beef Ranch

Larry, organic beef farmer

17

monuments to the lack of accountability of Alberta oil and gas developers.

Alberta law states that the corporations are responsible for maintaining payments over a lease's life, and restoring the site after operations cease. Pipelines, however, are not mandated for reclamation and may be left behind indefinitely. And the fickle nature of the market economy has made the terms of leases difficult for companies to keep — infrastructure constantly changes hands; companies go bankrupt and just can't pay, or are simply too small to foot the costly reclamation bill. Exemplifying the first of these dangers, Larry showed us a power line that veers across his land to service gas wells on other properties. The line's lease of $25 per pole per month hasn't been paid for over two years, ever since it was sold by the original corporate owner to Fortis Inc., a Canadian electric utility. Fortis says Choice Resources, the owner of the wells that the power lines service, is obligated to pay the lease. Choice Resources says it can't find the lease, so refuses to pay it.

The gas well on Larry's property has been defunct for over 10 years, and is surrounded by signs warning of poisonous gas. Although its lease was to be renegotiated by July 2006, no satisfactory agreement has yet been reached. Larry described his last round of negotiations:

"At the last meeting with the company landman I invited a number of Chinook Area Land Users Association members to attend; they all have leases and problems with the company. We have all since got letters from the company stating they will meet only with the landowner of that specific lease and no one else may attend . . . It's to keep one rancher from knowing what the other one gets and to prevent the sharing of information. Basically controlling information, they can tell each one a different story. . . . This is standard company practice — to pick people off one by one. We are preparing a response."

The two boards, the Surface Rights Board and the Energy and Utilities Board, responsible for hearing landowners' grievances, are "widely regarded by landowners as an arm of the oil companies," said Larry. And corporations are notorious for their tactics to delay expensive cleanups: they may continue to operate a well at minimal, unprofitable levels so that it won't qualify for reclamation, or they may stall the process by drawing out hearings and producing excess paperwork. Larry's well may join the thousands of other decommissioned but unreclaimed wells scattered throughout Alberta, and for now he has no choice but to coexist with this dangerous, corroding landmark.

But the intrusions of resource extraction are not limited to the comparative annoyance of bric-a-brac machinery, war zone–like poison gas hazards, toxic gunk and unpaid leases. Larry showed us an earthen platform on top of which sits a gas well. The well platform and access roads have disturbed nearly three hectares of Larry's land. During the platform's construction, the fertile topsoil was mixed with the clay subsoil, so it will take centuries for this site to recover its original fertility. In addition, the construction company planted crested wheatgrass on the disturbed site to prevent erosion. The wheatgrass — chosen for its drought resistance, because it establishes easily, and because the seed is cheap — does hold the soil in place and provide spring forage. But unlike the native fescue it replaces, it does not store up sufficient amounts of protein for cattle to over-winter on when it goes dormant in the fall, thus causing Larry the loss of a winter pasture. Also, crested wheatgrass has been shown to degrade soil over time because it concentrates its energy on above-ground growth while maintaining a relatively limited root system, the part of a grass that enriches soil with nutrients and organic matter. Over the years, the wheatgrass will migrate to adjacent lands, crowding out the native grasses severely impacting Larry's grass-fed cattle operation. Larry hesitates to mow or hay the pest grass because of the risk that his blades will strike scrap metal, left over from the well's construction and maintenance, hidden in the grass and spark off a prairie fire.

Other non-native plant species were introduced inadvertently during well and access-road construction, including houndstongue (*Cynoglossum officinale*) and Canada thistle (*Cirsium arvense*). Houndstongue is poisonous to livestock: Larry hasn't lost any animals to it, so far. But the thistle, introduced with gravel during construction of an access road, is the bigger problem. The weed is rapidly spreading across his property, displacing nutritious cattle forage by crowding out palatable native species. Bound by his organic licence, Larry is prohibited from combating the invasion with herbicides and instead has resorted to severing the flower heads by hand, a painstaking task in which he has invested over 200 hours of paid labour to date. The companies responsible have yet to compensate him.

Besides making it harder for him to make a living, by endangering his water supply, the drilling of oil and gas threatens Larry's very existence on the land. Gas and oil wells often traverse aquifers to reach deep fossil fuel deposits. Even properly drilled and maintained wells may lead to contamination of the aquifer from surface and subsurface sources — by the introduction of bacteria, by the downward seepage of surface pesticides or other foreign solutes, or by the upward migration of fossil fuels. Also, extracting fossil fuels changes the pressure in underground cavities and may alter the subterranean geography, causing water wells in the area to go dry. For example, drawing down gas from below the aquifer can result in the depletion of the aquifer as it drains to newly cleared lower levels. There are regulations that are intended to ensure that wells don't pollute the aquifer. But even if the well functions properly, the subterranean geology of aquifers and fuel deposits are rarely mapped in detail, making risks uncertain. Once contaminated, aquifers are virtually impossible to restore.

One might ask: what does this small ranch over 1,000 kilometres from Fort McMurray have to do with the tar sands? For me, Larry's ranch represents a test tube where corporations have been experimenting with fossil fuel development on a micro-scale, and the results of this experiment urge caution. Given that Larry's experience is the norm, not the exception, and given the corporations' obvious lack of stewardship and co-operation, should Albertans really be so eager to invite these same companies to bring their act to the Athabasca region on a mega-scale? In fact, current tar sand developments already display uncanny and telling parallels. Just as Larry's land has been recklessly subjected to weed invasions and topsoil degradation, with a total disregard for the future viability of the land, so also have

Walking at Bar U Ranch National Historic Site, Longview

the huge tracts of land in the tar sands and their exquisite ecosystems been irrecoverably laid waste. While Larry must live with the risk of water contamination and sour gas, the Athabasca River has already been found to be laced with heavy metals and carcinogens due to tar sands industry, endangering the health of the town of Fort Chipewyan downstream. While Larry's subsistence has been jeopardized by these intrusions, Aboriginal ways of life have already been disrupted forever by the pollution and massive destruction of habitat that is part of the development of the tar sands. From an economic point of view, just as Larry is being short-changed on his rightful lease fees, Albertans are letting their finite natural resources drain off to the U.S. while foreign corporations enjoy most of the windfall profits.

Still, resistance is growing, and there is no better example than that of Francis Gardner, a family rancher whom we met days later during our stay at Chain Lakes Provincial Park. Francis has a strong connection to the land, guarding it not only as his life, but with his life as well. He owns a ranch on the undulant Eastern Slopes

The group in front of the Gardner ranch

that was homesteaded over a century ago and has remained in his family ever since. His cattle help maintain the grassland, filling the niche once held by the now vanished buffalo. Someday, when he grows too old to carry on, he hopes to pass the ranch along to his daughter. However, not long ago his security on the land was imperilled by proposed oil drilling that would have endangered his aquifer and disturbed the grassland ecosystem.

Francis explained his responsibility to the land: "What you see now around you is a reflection of the past and what you see a hundred years from now will be a reflection of what we do today. So if we cut it up and put acreages on it and oil wells and roads and all the rest, this country will not be here as you see it today. It will be something else and it will be permanently changed forever. The water in the aquifers, water in the streams — that's what keeps southern Alberta watered, and if they mess that up, the whole country's going to pay a pretty mean price for it."

When he found out about the drilling proposals, he fought through the bureaucratic system, and lost. Soon oilmen began arriving at his ranch, demanding entry for seismic testing.

Francis blocked their way, denying them access by ground. Not long after, helicopters arrived on his property carrying seismic equipment. He put his body on the line by standing beneath the helicopters with a group of neighbours to prevent them from landing. "They use brinkmanship to get their program in place and they'll play dirty tricks on you if you go against what they have in mind," Francis explained.

For now the oil company in question, Shell, has relented in its drilling attempts, wishing to avoid confrontation. However, no binding agreement has been reached between landowners and Shell, so, as the law stands, Shell could push ahead with drilling and exploration plans. In the meantime, Petro-Canada has announced plans to force a pipeline through 15 miles of virgin wilderness to the west of Francis's land. Francis concludes, "This is the fight now . . . it never ends."

– *Greg*

101 Variations of Chili

Kealan and I stayed longer than we had intended at the Diamond Willow Ranch enjoying our steak lunch and playing music with Larry Firth. By the time we were back on the road, the day was leaning towards night and the rest of the bikers had undoubtedly already arrived in Pincher Creek. We envisioned them gaunt-faced around a boiling pot of water, arguing over how best to cook us when we slunk in late with the group's food. To complicate matters, our group meals consisted largely of grains and dried foods that required considerable preparation time, a fact that would extend our companions' hunger and prolong the scorn we would have to endure while waiting for their rice to cook. Weighed down by our own full stomachs, we gasped and heaved, making up lost time as best we could.

Suddenly Kealan, who was riding in back, yelled for me to wait. He spun his bike around and stooped to pick something off the shoulder of the highway. Raising a bag over his head he let out a victorious whoop, "FRESH PIZZAAAAA!" and I rode closer for a better look. To my amazement, he was right: like manna from heaven, the extra-large zip-lock bag contained the intact remains of most of a pizza, presumably fallen from the roof of a car. It was as if God had intervened on our behalf, sparing us from the wrath of our ravenous group mates!

After our initial jubilation, however, it sank in that our predicament was not yet resolved. God's goods "hot off the asphalt," so to speak, might seem a dubious blessing to our friends. We carefully planned our delivery, deciding to tell them only that the pizza had been a "gift" from a sympathizer, a partial truth under the circumstances. Surely our benevolent God would forgive this fib which He Himself had obliged us to tell.

Finally at camp, eager to allay the hunger of our friends (and turn away their wrath), I hurriedly passed out the pizza. But as I was about to tell the euphemistic story of its origin, my conscience won me over and I came out with the truth. "We found it on the road," I said in a confessional tone. Aftab, with eyes only for the pizza, seemed to ignore me and impatiently held out her hand to receive her piece. Well, apparently our group is much less squeamish than I thought, I told myself, and here is proof that there is no reason to be afraid of telling the truth.

After dinner, hunger silenced, Aftab became more curious. "So really, where did you guys get that pizza?" she asked. Kealan assured her that we were telling the truth, which she brushed off. "Enough with the jokes; that was good pizza — where did you get it?!" I tried again, "God dropped it off, manna from heaven," but only succeeded in provoking her more. It took some more convincing, but with much groaning and laughter our friends finally believed us.

The special delivery at Pincher Creek was a

Our evening meal preparation sounded something like this:

Kalin: "We have rice!"

Me: "Put it in the chili!"

Shawn: "We have lentils!"

Me: "Put it in the chili!"

Jackie: "We have chocolate chips!... oh no you don't, Greg!"

Oatmeal, we found, could be innovated with similarly satisfying results.

— Greg

one-time gift, and our fare relapsed into oatmeal breakfasts and chili dinners with leftovers from one or the other as lunch, day after day. Monotonous as they sound, these would turn out to be some of the most satisfying meals I have ever eaten. One needs only to realize that chili is an infinitely variable dish, since its name refers to only one ingredient: chili powder. Taking advantage of this loose definition, we pushed the boundaries of known chili recipes, adding everything from squash to pasta to our concoctions.,

As one might expect, our mad experimentation soon got out of hand, and the group had to all but restrain me by force from experimenting with the ultimate fusion, adding oatmeal to the chili, and thereby destroying the boundary between breakfast foods and dinner foods that maintains order in the universe.

Finally, at Lac La Biche, we had had enough. Someone proposed dumpster diving and the group responded with overwhelming enthusi-

asm. Interspersed with the odd banquet, compliments of one or the other of our gracious hosts, oatmeal and chili had run their alternating course through our digestive systems like concrete and battery acid, so one might understand our excitement at the prospect of eating from the garbage. Had we needed them, our meagre food budget and philosophies about waste were but further reasons to "dive." Eating from the garbage, however, by no means implies eating garbage. I have scrounged many a (to my eye) flawless apple from grocery dumpsters in my

Dumpster diving

in the centre of our unorthodox food encounters, stood tall inside the green steel container, plunging out of sight and resurfacing, arms filled with the bounty of expired products and lightly bruised fruit. It was all that Shawn and Dylan could do to keep up, relieving him of his winnings. The spree came to a disgusting halt when Kealan yelled, "Ugh, broken eggs." There were rustlings from the dumpster, then a muffled "Yuck! Rotten meat!" Finally, Kealan sprang out of the dumpster, a single orange clasped in his filthy hands. Doing a victory dance, he superficially wiped it on his pants and proceeded to peel it. I watched, too appalled to speak, as he bit into the half-peeled fruit before generously extending a rot-stained hand to share with the rest of us. Noticing the looks of revulsion on our faces, he asked "What?" with total innocence, then noticed his hands and wiped them on his pants again. He proceeded to eat the rest of the orange while we laughed with horror. To our great shock and relief, he did not even contract a runny nose from this egregious lapse in hygiene.

The boxes of food, including pastries, breads of all kinds, tomatoes, parsley, zucchini, squash, an entire box of apples, carrots and a half-full bucket of muffin mix from the bakery, filled our economy car like a cornucopia. Once at camp we enlisted all hands, deciding that the food must all be cooked to make sure it was safe, even the fruit. Soon applesauce was simmering in its pot on a borrowed barbeque grill, and muffin-pancakes were being flipped in the skillet over our campfire. Two casseroles were prepared from the vegetables, and there was bread for all. In the end, the results of our scavenging fed 18 of us for three days.

—Greg

home town. Conversely, I have eaten worse from my own college refrigerator. So with much gusto, Kealan, Shawn, Dylan and I hit the alleyways for some good modern-day hunting and gathering.

I became separated from the group in the frenzy of the hunt, and when I found my cohorts they were unloading loaves of bread by the crate from a grocery store dumpster. Kealan, as usual

Aftab and Dylan approach a turbine

August 17

Harnessing the Elements in Pincher Creek

As we pulled into Pincher Creek on that first trying day, our chapped lips and reddened noses proved that the wind blows strong and the sun shines hard in southern Alberta. The power of the elements is not lost on the residents here. Wind turbines are as much a part of the landscape and local culture as the grazing cows and rolling fields of barley. The road sign welcoming visitors to Pincher Creek boasts that "Our hospitality will blow you away." Solar panels are less symbolic of the region's commitment to renewable energy, but there is lots of talk about the potential of the solar industry as well.

The rule in the wind industry, as with most things in Alberta, seems to be "go big or go home." Large farms with hundreds of turbines are the norm, but according to Hal Jorgensen, Operations Supervisor at TransAlta Wind in Pincher Creek, a certain threshold has been reached: "The transmission line is full. Until we get new transmission lines, there will not be new building," explained Hal.

While construction and maintenance are supplied locally, the manufacturing of towers and turbines is still largely controlled by the European market. TransAlta Wind buys its turbines from Vesta, a Danish-owned company. Meanwhile, plans for a local manufacturing industry appear to be at a standstill.

"I really don't know what's in the queue with that," said Hal. "I know there's been talk. I think that that requires some level of government helping out. I know that the Quebec government has been quite involved in trying to bring manufacturing to that area. Quebec has some tower manufacturing and there is talk of blade manufacturing. The Alberta government has talked about manufacturing, but there's been nothing so far."

It was unclear from my conversation with Hal what incentives, if any, exist in Alberta for the production of renewable energy. Other jurisdictions, such as the Province of Ontario, have established

Jackie, Dylan and I decided to take what we thought was a 10-kilometre detour (turned out to be 40 kilometres!) to get close to the wind turbines. At one point we were riding on a dirt road with massive turbines against a red stormy sky on one side, and on the other side cows — stopping to watch us go by — under beautiful blue skies with fluffy white clouds and the rays of sun shining down like God is sitting up there watching! My new image of paradise.

— Aftab

Standard Offer Programs, guaranteeing a fixed price per kilowatt hour for renewables over a fixed period of time. There is no limit to the number of projects that may apply for contracts, but the size of each project is capped at 10MW (megawatts). The idea is to decentralize power generation. This approach to energy production is often referred to as Community Power — keeping energy dollars at home and opening up the renewable market to small-scale producers. Alberta, the land of free enterprise, is far from adopting this approach.

"In Alberta, the energy market is deregulated," explained Hal. "As far as I know, there's not a lot of preference given to whether you are coal, natural gas, wind, hydro. It's basically if you want to build it, great. We do get some government support, but of course Alberta is known for its oil and gas. There certainly aren't the profit margins in the electrical industry that there are in oil and gas. That's pretty obvious to everybody."

So are the tar sands stifling the growth of the wind industry?

Hal believes so: "I'd say the way that it affects us the biggest is wages. For instance, if a guy can make 50 bucks an hour in Fort McMurray, he's not going to want to work down here for 15, so it certainly does drive up the wages. We are actually starting to experience labour shortages down here as well."

The logos on the service trucks at TransAlta Wind still read Vision Quest, the name of the start-up company, recently bought out by the larger power generation and wholesale marketing company TransAlta. Vision quests are a traditional practice of the Confederacy Blackfoot who have for centuries ventured alone into the surrounding hills in search of a spirit companion to guide their lives. The rise of wind power as a serious player in the world energy market proves that alternatives to large-scale energy projects do indeed exist, but my conversation with Hal made me realize that the wind industry has been unable to escape the forces of globalization and capitalist economics. The industry needs guidance in order to expand sustainably.

The route out of Pincher Creek took us past the myriad of turbines we'd been admiring up on the horizon. Aftab remembers that day as the most beautiful day of the bike trip.

We had a good chat that day about the visual appeal of the wind turbines. This is one of those things society is pretty divided on. I clearly felt the turbines were amazingly beautiful, but was that because of the connotation of clean energy that I hold so dear? Do massive structures like that only look attractive if you believe in their goodness? I was somewhat alarmed at the level of dislike for wind turbines in Pincher Creek among the residents. Even one old-time environmentalist we spoke to complained about their presence and how they are disturbing the landscape. For me, you have to put it in context. Given that we need energy to sustain our quality of life, would you rather see wind turbines out your window, or the smokestacks of coal plants?

— Tim

Tonight's Digs: Camping on Pincher Creek, right on the creek that gave the town its name.

We arrived at Pincher Creek during Rodeo weekend!!!

August 18
Riding in the Rodeo

As outsiders trying to capture local perspectives on a provincial (and national) issue, it was important that we allow ourselves to see things from conflicting points of view: the vegan among us must try to understand the appeal of the rodeo, the indie rock kid must find an appreciation for country music, the hippie environmentalist must empathize with the migrant oil worker, the revolutionary must find commonalities with the housewife in Fort McMurray. The flight attendant, meanwhile, pedalled along questioning her own oil-dependent profession. We bikers were a diverse cast of characters to say the least.

We arrived in Pincher Creek during Rodeo weekend. For most of us, a largely urban and vegetarian crowd, this would be a first foray into the cowboy lifestyle. Could there be anything more quintessentially Albertan than a rodeo? We couldn't have written a better script. We had hoped to get a cross-section of Alberta and we were getting it. One of the rodeo events we watched was bull riding. An enormous bull, about 800 pounds of pure energy, is caged in a pen just

Gorging on pancakes from the giant griddle!

long enough for the cowboy to drop onto its back and grab the manila rope around its middle, his only handhold. A rodeo clown jumps around, getting the bull agitated so that when it's released, it is bucking wildly, then the clown dives for cover into a barrel in the ring. The cowboy holds on for as long as he can, maybe 12 seconds.

The next morning, several of us got up early, made our way to the rodeo pancake breakfast, and gorged on griddle cakes and sausages. It's difficult not to stick out in a small town like Pincher Creek, especially when you're dressed in spandex. As we cycled back to our campground, each with one hand on the handlebars and the other holding a stack of pancakes for the late risers among us, we caught the gaze of the many town folk beginning to line the streets with lawn chairs for the annual rodeo parade, in which, you guessed it, we'd been invited to participate.

Punctuality was not a strength of our group, nor of any group I've ever been a part of. If group touring has taught me one thing, it is patience.

You're always waiting for something or someone. Dylan is making himself a sandwich. Jackie just got a flat. Aftab is taking a call from a reporter. Jodie is struggling to strap her camera onto her rack. Kalin is proposing a quick meeting. Shawn is . . . Where *is* Shawn? Has anyone seen Shawn?

And so, as usual, we were late for the parade. A few of us bolted ahead to grab a spot towards the end of the procession, while the rest straggled behind at the campground. As chance would have it, we stuck ourselves right behind a local politician pulled by horse and buggy. There was a smell of horseshit in the air. In fact, this parade had more horses than any I'd ever seen. Horses, then bikes, then automobiles. That was the order of things, the evolution of transportation. Our appearance in the parade was rendered somewhat anticlimactic as the stragglers meandered their way through the crowds now lining the main street, and up against the flow of people now parading forward, to meet us at the back of the pack. But the campaign machine was in full gear. Sporting our www.tothetarsands.ca

The ride into Fort MacLeod was a long one for many of us. As we made our way out of Pincher Creek shortly after noon, the source of the mysterious clicking noise coming from my bike's drivetrain that had been bugging me since Waterton made itself known. My freewheel came loose and its bearings went flying. Unequipped to rebuild the wheel myself, I took to the side of the road and attempted to hitchhike to the nearest bike shop, which was in Lethbridge, over 100 kilometres away, in the wrong direction. Several cars pulled over, but all were headed to Piikani, a Native reserve of the Blackfoot Confederacy, a mere five or 10 kilometres up the road.

— Tim

tar was able to come along too, but there were times when it was less welcome. A few days later camping in Chain Lakes Provincial Park our early morning wake-up call was Kealan belting out Johnny Cash's song about death row, "I've Got 25 Minutes to Go!")

The Piikani Reserve is Alberta's second largest in area, but like most reserves it is visibly poor. The reserve is located on a great plain in the foothills just north of Pincher Creek. The Pincher Creek turbines are visible in the distance, but Piikani is situated just beyond the prime sites for wind development and so the community has missed out on the land-leasing opportunities that come with it.

After declining the first few rides in the hope of getting a lift straight through to Lethbridge, I got in with Vince, throwing my bike in the back of his pick-up and squeezing into the cab up front. Vince offered to take me to his home on the reserve, where I'd be welcome to rummage through his son's pile of scrap parts. He was sure I'd find a wheel in there, and ultimately I did.

On the ride back to the highway from the house, Vince spoke of his experience as a modern Native man. According to Vince the recent popularization of pow-wows, dream catchers and

T-shirts and banner, we walked next to our heavily loaded bikes, passing out pins and waving like the Queen as Kealan played his guitar and sang:

> When mangoes cost 300 dollars
> Will you still commute?
> Why wait for then?
> The time is now
> Come on everyone, let's bike!

(Kealan is so tall and lean that he was thrilled to have found a bike with a big enough frame to take him, and his guitar, across the prairies. For the most part the rest of us were thrilled that the gui-

Native clothing has cheapened the culture by undermining its spiritual components. He told me the story of Head-Smashed-In Buffalo Jump, a nearby historic site. The Blackfoot hunted buffalo there by driving them off the 10-metre cliff. According to legend, a young Blackfoot wanted to watch the buffalo plunge off the cliff from below, and was buried underneath the falling buffalo. He was later found dead under the pile of carcasses, "his head smashed in."

That afternoon, as I rode alone towards Fort MacLeod, chased by an ominous-looking rainstorm, I reflected on our chaotic, yet fruitful first few days, my frustrating morning of roadside repair and my unexpected encounter with Vince. Then and there, I decided on a road philosophy for the weeks to come: Expect nothing and be prepared for anything; abandon oneself to the experience, see misfortunes and obstacles as opportunities for adventure, and open one's mind and body to the world of possibilities.

– Tim

Tonight's Digs: Dutch couple's backyard in Fort MacLeod.

August 18
The Security and Prosperity Partnership

We met Sheila Rogers and her husband at a gas station on the outskirts of Fort MacLeod. Sheila works for the Public Interest Research Group (PIRG) in neighbouring Lethbridge. She brought us to a nearby park for a picnic starring her delicious homemade chili. We arrived in Fort MacLeod just as Prime Minister Stephen Harper was set to meet with U.S. President George W. Bush and Mexican President Felipe Calderón in Montebello, Quebec, to continue their closed-door discussions on the proposed Security and Prosperity Partnership (SPP) of North America.

As part of its campaign Integrate This, the Council of Canadians was calling for a national day of action in opposition to the SPP. We'd answered the call and were looking to make issues of free trade and energy sovereignty a focus on this early leg of the tour. Sheila summed up the SPP as follows:

"The SPP is very undemocratic. It completely circumvents the entire democratic process because the citizens of Canada have not been consulted and not even Parliament has had a chance to debate it. All the shots are being called by CEOs of major corporations . . . whose only responsibility is to make profit for their shareholders. They don't care about citizens and their right to water and their right to sustainable energy; they just care about making profit, and they're the ones that are making these decisions that will affect Canadians and North Americans all over.

"The purpose of the Security and Prosperity Partnership is economic and political integration of North America. We're trying to harmonize all of our rules and regulations. One of the reasons for the SPP is that the U.S. wants to secure their own energy and water supply for their own people, and since they don't have enough in their own country, they're looking to Canada to supply both water and energy for them. Under NAFTA, once you export a certain percentage of energy, you can never export a smaller percentage than that. It always has to be an equal percentage or more. The U.S. is encouraging the oil sands (of Alberta) to develop and export fivefold what they are exporting now. In fact, they're building a pipeline in Alberta right now [Keystone Pipeline] that will take the raw bitumen from the tar sands to the U.S. Not only are we losing the energy, but we're losing jobs that could be had in processing this bitumen. The SPP is just fuelling the whole fire that's going on in the tar sands right now and making it so that it'll never be able to stop, it'll just keep on going; so that's what we're concerned about here in Alberta."

Always the devil's advocate, Kealan would challenge us to consider the other side's perspective. He asked Sheila why she thought Canadians should hoard their resources when the majority of the demand for them comes the United States. She responded that she thought it was "really important that we secure our own energy and water needs first before we start sharing them with other people and make sure that we're not giving so much away that there's not going to be enough for us. As long as we have an actual plan, a national energy plan, and know what we're doing, then if we had extra, we could share it and we could sell it and make money, and the same with our water, but until we have our own needs secured I don't think we should be sharing it with anyone else."

So do we have a plan? Do the Americans? Can a government that is willing to allow a full third of a province to be sacrificed in the name of progress and development pretend to have any real semblance of a plan? Can a society that continues to build casinos and golf courses in what are essentially deserts be said to have a serious plan for the efficient use of its natural resources?

From all indications, we have lost our energy sovereignty and we are without a plan to get it back. *Fuelling Fortress America*, a joint report by the Canadian Centre for Policy Alternatives and the Polaris and Parkland institutes, calls for "a made-in-Canada national energy policy and strategy for the twenty-first century." The report states that "there was a time when Canada was somewhat sheltered from the international oil crises. . . . During the 1970s and up

Our party in the park with Sheila Rogers

Leaving ranchland behind us

heat 1.5 homes for a day are used to extract the bitumen. Should we not instead use these resources to manufacture wind turbines and solar panels, to heat our homes, to rebuild our rail network and, in the case of water, to simply quench our thirst?

So while Harper met in private with Bush and Calderón, taking cues on energy policy from the heads of North America's largest corporations, we sat in a public park and openly debated the merits of resource sovereignty until the sun began to set on Fort MacLeod, the self-proclaimed sunniest town in Canada.

— *Tim*

August 19
Human Power and Oil Power

Before I signed up for the To the Tar Sands bike trip, I had to ask one question: are we going to have to ride in the Rockies!? I am a city biker, you see. Typically, I ride to get to work and get to dance class and get to the bar. The only other major bike trip I had done was 10 days in Holland, which is famous for being flat and made for cycling. I will be honest: I was relieved we didn't have to cross the Rockies. The rest of the prairies are flat, right? Right.

On the third day of trip, we had our first "100+ km" (one hundred plus) day riding from Fort MacLeod to Chain Lakes Provincial Park. I later came to know only two types of days: the less than 100 km days and the 100+ km days. And I learned not to trust the more precise estimates of how far we actually had to go. Despite our advanced Google-mapping skills, supplemented by more primitive techniques (one day we used a blade of grass to figure out the distance on the map from a poorly drawn scale bar), we never truly knew how far we had to travel each

to the 1980s, Canada maintained a two-price policy for oil — a domestic price and an international price — which meant that the international increases by the OPEC nations did not affect Canadians. That favourable policy for domestic consumers was cancelled when the Mulroney government repealed the National Energy Program in 1985."

The world has reached peak oil and as a result production of oil from the tar sands is finally economically feasible. If we acknowledge that oil and gas are finite, even if they are still plentiful here in Canada, should we not then immediately begin to reorient society to make more strategic use of the precious little oil we do have? Consider that to produce one barrel of oil, four tonnes of material are mined, and then two to five barrels of water and enough natural gas to

Camping at Chain Lakes

water, as places to pee — given the nature of our bike ride; but that's a different story.) To this day I am convinced there was no gas station on the right-hand side in that little town. To this day, Greg is convinced he waved to me violently from the side of the road, but I ignored him and biked right by. I guess we'll never settle that one.

At any rate, I was by myself half an hour outside the town and I noticed it was getting hilly. I stopped on the side of the road and

day. Basically, the actual length of the trip was guaranteed to be longer than the calculations suggested.

This original 100+ km day was particularly special because it was also the hilliest day of the trip. We were climbing up into Chain Lakes Provincial Park and the locals had told us there would be a big hill. One big hill. Not a big deal.

I remember I was separated from the group that day because I never managed to find our lunch meeting spot. We were supposed to stop at the first gas station on our right in a small town before climbing the big hill. (I always thought it was ironic how much we depended on the existence of gas stations — as meeting places, as places to get

There had been a big debate on our listserv before the trip on whether or not we should have a support vehicle. On one hand, the point of doing this trip by bike was to do it sustainably and without fossil fuels, the merits of which we were questioning by the very fact that we were in Alberta. On the other hand, we needed to ensure the group was safe and able to get from one place to another in a reasonable amount of time, so that we could achieve our objective of meeting with Albertans and hearing their stories. The night before the long ride to Chain Lakes, we had come to some form of consensus that we needed a car, if only to carry the food and cooking supplies that were weighing us down. At some point on my solo ride into the park, Kalin showed up driving the newly rented car (oh, how shiny it was on that first day!) and relieved me of some of the weight on my bike. I had never been so glad to see Kalin!

— Aftab

ate a huge pile of dates and walnuts as my lunch. I gave a quick call to my sister in Vancouver because I had a song of hers stuck in my head but I couldn't remember all the words and it was driving me crazy! Then I called the rest of the bikers to let them know I was ahead. Jackie asked me if I had passed the big hill yet. "Well, I have passed a bunch of hills. I am not sure which one was *the* one."

As it turned out there was no one big hill, but there were many, many, many to follow! One after another and steep as can be. I must have been climbing hills for hours, stopping at the top of each for a drink of water. After a while I stopped counting and I stopped wondering if or when I would arrive. I had been leafing through a book by Pema Chodron on the basics of Buddhist meditation earlier that day, and on those steep hills the meditation lessons began to take on a new meaning:

"Okay, so let's face the facts: I am in pain, my legs are killing me, my back feels like it's going to break in half, I am hungry, I can't keep myself hydrated for more than five minutes, I am alone on these huge hills in the middle of nowhere! And you know what? I can't change any of that! So I might as well start to relax my fingers on the handlebars and my toes on the pedals. I might as well start to enjoy the wind that's touching my face. I might as well start to see the beauty of Alberta's endless skies. I might as well notice the fascinating shaping and colours in the asphalt. I might as well say hello to the llama waiting for me curiously at the fence. I might as well accept my situation, stop blaming myself for not being stronger, and see what the next moment brings me."

I remember that ride as a sort of spiritual awakening. It certainly brought on a change of attitude for me. The trip was going to be difficult, but there were ways to be okay with that. There were even ways to enjoy it.

On the same day we biked into Chain Lakes Provincial Park, we also got ourselves a car. And so began a very special relationship between the bikers and the support vehicle. It remained a love-hate relationship for many of us to the bitter end. The car introduced a variety of new roles into our group dynamics:

The role of support and the role of speed — The car was able to cover, in about half an hour, the entire distance that we were able to bike in a typical day! This meant that the car could go back and pick up stragglers, of which there were often several in one day. It could go ahead of the crowd and find us a place to sleep or get us food to eat. It could be a temporary roof over our heads if we were sick. It could rush someone to an interview when we were late. (One day in Fort McMurray, it would allow six of us to make it to our little flight over the tar sands, less than 10 minutes before it was scheduled to depart!) The car was very practical.

The role of fair share, and unfair share — The role of the car introduced the role of the driver. The role of the driver wasn't a particularly popular one, because the driver would miss out on much of what happened during the day, and on top of that often had to clean up after the rest of us and pack and unpack an impossibly full car sometimes several times a day! As it turned out, only about half of us on the trip actually knew how to drive, so the role of the driver couldn't be shared equally among the group. The truth is, we weren't all equal on the trip. Even if we wanted to be very fair about everything and share all responsibilities equally, that it wasn't practical was highlighted by the car. Our view on fair and unfair share had to change.

The role of additional adventures — The car came with its own package of adventures we hadn't anticipated. The most famous of these adventures was when Jackie locked the keys in it on the side of the highway. I remember pulling up to the car with Dylan and Greg, while she was being assisted by a young couple using a hanger

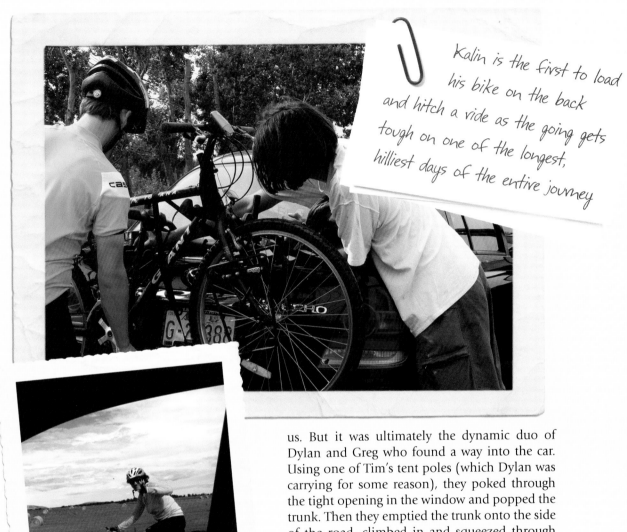

Kalin is the first to load his bike on the back and hitch a ride as the going gets tough on one of the longest, hilliest days of the entire journey

us. But it was ultimately the dynamic duo of Dylan and Greg who found a way into the car. Using one of Tim's tent poles (which Dylan was carrying for some reason), they poked through the tight opening in the window and popped the trunk. Then they emptied the trunk onto the side of the road, climbed in and squeezed through the back seat to reach the door lock and open it! Later Greg formally thanked Jackie for locking the keys in the car and offering us the opportunity to feel as creative as we did!

The role of modern comfort and convenience — Sometimes the car felt like the only device attaching us to the conveniences of the Western modern world. I remember the first time I climbed into it. It smelled like a man in a suit — cologned and clean, and probably somewhat

to try to open the door. (It was only our second day with the car and the last day of its scratch-free existence!) One great thing about Alberta is how helpful people are. Many stopped to help toxic. It brought up memories of being new in Canada and experiencing the luxury, warmth and softness of Western life. Of course, the car has become a symbol for modern convenience — a lifestyle that depends on oil. The car played the role of reminding us of that lifestyle, which is actually the lifestyle many of us live when we are not biking to the tar sands. The majority of our lives are spent in a familiar comfort that links us directly with oil. We are players in the drama of an oil-dependent world.

The bikers and the car. The human-powered and the fossil fuel—powered. At least we were clear that the second was only there as a backup to the first. Something to think about as we coasted down the hills out of Chain Lakes and towards our next stop: Turner Valley.

– Aftab

Tonight's Digs: Chain Lake campsite on top of impossible hills, no water or electricity.

August 20
The Remnants of Alberta's First Big Oil Discovery

I have been visiting Turner Valley my whole life. It's a small place (pop. 1,800) with one major intersection, a gas station, and a bit of an empty feel. But the people are friendly, the hotel bar has great music on the weekends, and from the outside, it appears to be a content, picturesque community nestled in the foothills of the majestic Rocky Mountains. And that's how I always saw it, until the trip . . .

It turns out that it is the birthplace of Alberta's oil and gas industry (the first major well was discovered there in 1914), and once the largest oil producer in the British Empire. Only a few hundred metres from the centre of town are the remains of the Old Gas Plant, now a National

Historic Site, closed while a government-appointed panel decides whether or not it should become a tourist destination with an interpretive centre. It sits on the banks of the Sheep River. The Oil Field Society arranged a tour for us, led by historian and author David Finch. Dressed as an oil worker from the forties, he gave us the history of the Old Gas Plant.

We gazed at the remains from across the river. I absent-mindedly assumed that the current decontamination work was reliable. All those stories of dangerous working conditions and crude and benzene bubbling into the river belonged to a different world, long ago. Native people had used the crude to waterproof their canoes for centuries.

That night we stayed at the Oil Fields Arena (in Black Diamond, a community three kilometres downstream from Turner Valley), courtesy of the Black Diamond Boys and Girls Club. Hot chili was prepared, and it was the first night we'd spent indoors after nearly a week of riding. We had Internet; some people brought us beer . . .

And then Roxanne showed up. I'd been really excited about meeting her ever since I'd heard she was one of the local activists responsible for the government spending millions on building a containment wall and monitoring system at the Old Gas Plant site. The government had purchased it for a dollar in 1988 and let it sit leaking into the Sheep River until it was designated a historic site and cleanup finally began, many years later.

We quickly left the arena and stood outside the back doors to talk. Roxanne is not always well received around town. The mayor of Black Diamond, David Finch and various other prominent people from the town were inside the arena, and she preferred to avoid confrontation.

She had a contagious nervous energy that hit me hard, as did the story she had to tell. She described flooding they'd had two years before during which soil from under the Old Gas Plant slumped into the Sheep River. There were three floods in three weeks, and they were higher than the flood of 1995 during which the Sheep River overflowed its banks. The contamination at the Old Gas Plant is considered Tier 3C. Jodie, who was familiar with Alberta's remediation guidelines for former oil and gas sites, questioned whether this legal category existed. Tiers refer to whether the site should be evaluated according to certain parameters for levels (that is, hydrocarbons cannot exceed so many milligrams per kilogram) or according to a more site-specific approach. The Old Gas Plant is in an area where there are naturally occurring high levels of hydrocarbons. Tier 3C basically means "forget cleaning it up — just take samples at a distance and try to contain it." However, this was a popular swimming hole for locals, and since people around here live to be 100 it was thought there was no problem.

So within a few minutes, I had gotten our water management scholar (Katherine, who'd joined our group only hours earlier) and documentary filmmaker (Jodie) and we were following Roxanne down the highway to go and investigate a flare that would be considered illegal by today's standards. It was getting dark as we pulled up to the banks of the Sheep River, next to the Old Gas Plant. Sure enough, there was a huge sour gas flare blasting out of cracks in the rocks. There was a fence around it and an electric ignition device making sure the flame did not go out, but given the well-publicized extreme fire-risk conditions at the time, I was a little surprised to see it at ground level, surrounded by dry grass and giving off a hell of a smell.

Less than 100 feet away and directly downhill from the tanks that held benzene during the founding of the oil industry was a drinking-water

We stayed in the Oilfields Regional Arena that night, and hung out in a closed bar to use the free wireless and plan the next leg

David points out the old plant

Aftab sketches the old gas plant

well for the town of Turner Valley. The well was almost level with the Sheep River but had three-foot-high walls of sandbags around it to protect it (which incidentally did not protect it during the flood in 2005). Roxanne has been trying to expand the parameters of the usual water testing to look for more kinds of contamination in Turner Valley's water supply, but has met heavy resistance.

The town of Turner Valley is building a reservoir on a piece of land just west of the Old Gas Plant site. Another concerned citizen had forced the town's lawyer to write a 72-page document to prevent Roxanne from having intervener status at an upcoming hearing regarding the concerns about the reservoir site. Many people in the town blame Roxanne for escalating taxes due to additional water-testing costs and legal fees associated with the town defending their decision to build a drinking water reservoir on a contaminated piece of land. But at $150 to $300 per hour for a lawyer, 72 pages of verbal diarrhea is definitely a waste of taxpayers' money.

Despite the Alberta Environmental Protection Act encouraging citizens to get involved with protecting the environment, it is not an easy process. Through the hearing process, citizens can submit personal costs for the work they contribute to the Alberta Environmental Appeals Board that helps them make their recommendations to the Minister of Environment. But Roxanne doesn't expect much reimbursement for the time she has spent doing research and attending meetings, all on her own accord. If someone were able to calculate the cost of future health care and suffering, plus the cost of the loss of the integrity of the water system in Turner Valley, a few billion now to really clean up the Old Gas Plant site and the rest of the town site would probably lead to significant savings.

We marvelled at the new containment wall between the Old Gas Plant and the Sheep River. It stops liquid drainage off the site and directs it downhill to a treatment facility. It was built as a result of Roxanne and her friend appealing to the federal government (Department of Fisheries and Oceans and Environment Canada through the Deputy Auditor General of Canada), which initiated an investigation. Another reason the need was recognized was that Roxanne was able to get the media out to witness the contaminated banks slumping into the river during a flood. The government doesn't like bad press.

Roxanne feels that the risk-management

Sour gas flare!!!

approach to dealing with the contamination at the Old Gas Plant is just ensuring future generations will have to deal with it and that the taxpayers of Alberta will have to continue paying for its upkeep (monitoring and maintaining it). "Budget cuts" was the excuse the government gave for doing nothing for 20 years, allowing the contamination to flow into the Sheep River. There are no guarantees that this government or future governments won't use the same excuses. Alberta has very stringent laws regarding the environment, but it appears that these laws are not being enforced.

– Kealan

Tonight's Digs: Oil Fields Arena in Black Diamond.

August 21

Okotoks — A City Recognizes the Limits to Growth

One reason Alberta is so interesting is that it hosts not only some of the most worrisome symbols of our modern unsustainable ways, but also some of the best solutions. Examples of individuals, companies and public agencies doing innovative work in the field of sustainability are scattered throughout Alberta. We got our first glimpse of these efforts at the wind energy fields in Pincher Creek. More examples of pioneering sustainability work awaited us as we left Turner Valley for the town of Okotoks.

As an urban planner, I have enormous faith in the power of local governments to save the world from many of the environmental and social disasters we have brought upon ourselves. The level of government I trust is the municipal. Municipalities are close to the ground, close to the people, and their decisions actually impact a vast majority of what we experience in our day-to-day lives, such as accessibility and quality of housing, transportation and green spaces. I was particularly stoked to get to Okotoks, which is the poster child of municipal sustainability in Canada. When we sat down with the town manager and elected officials in Okotoks, I could see why. The town of Okotoks is the fastest-growing mid-sized town in Canada and the nation's second-youngest community. It primarily accommodates people who work in Calgary, a half-hour drive away. It seems as though Okotoks could grow indefinitely. Instead, the town has made a unique and interesting choice, which speaks directly to sustainability. It has made the conscious decision to grow only so far as the land can support the local population. In other words, the town has set an upper limit of 30,000 people, reasoning that this number can be accommodated within the carrying capacity

of the town's geographic boundaries. The most important consideration of carrying capacity in this case is water. As we learned quickly on our journey, water is a huge issue everywhere in Alberta. It is one that governments, industry, environmentalists and citizens are all concerned about. The Town of Okotoks has decided that it will not import water to support its people, but will instead take the self-sufficiency route. This means that significant water efficiency gains must be made, so that the limited amount of water available within the municipal boundaries can accommodate a growing population.

Other measures are also in place to protect watersheds. For example, topsoil in some of the newer residential developments must meet a minimum standard depth of eight inches. The increased permeability of the soil decreases runoff during times of melting and heavy rains, while enhancing the natural filtration process. This speaks to the invaluable ecosystem services provided by our often neglected soils. In any case, this particular measure was especially encouraging to me, for it is more about intelligent redesign and mimicking of nature than it is a technological quick fix.

Okotoks is also famous for its approach to renewable energy. The town itself supplies most of its electricity from renewable sources, but the project that has made national and international headlines is the solar community of Drake Landing. Somehow, Shawn had arranged for us to actually stay overnight in the solar community, where we were hosted by the kind homeowners. At first glance, Drake Landing is not so different from any other suburban subdivision. And that is exactly the idea behind this pilot project: "Let's make a community that is in every way representative of the typical suburban North American subdivision — except that 95 percent of its energy needs are supplied by solar power." You can look closely and see the

A typical solar home in Drake Landing

Cookies and juice upon our arrival

The mayor tells us his vision

On our way to Okotoks, we stopped at the Big Rock for a photo op with local media

solar panels covering the roofs of most buildings, or check out the impressive control panels and meters in the basement of every home that keep track of how much power is generated and used. Drake Landing proves a point: our modern housing does not have to depend on fossil fuels.

The sustainable technologies employed at Drake Landing were impressive, yet the place lacked a certain aesthetic. Most of the homes were still much too large for the number of people who inhabited them. Architecturally, they followed a certain cookie-cutter pattern. While the innovations around energy were top of the line, many of the more basic sustainable design practices seemed to be lacking. During our tour of the expensive solar collector station, Tim asked whether the homes had been oriented and designed for passive solar. They had not. The main intent of the project (and the attached government grant) was to experiment with new forms of solar technologies. Fair enough, but far from the ideal community I would someday like to live in.

That night, our host families treated us to a fantastic potluck meal cooked on high-efficiency stoves. Despite the manufactured feel of the neighbourhood, community was forming here organically. Later, we rinsed off under solar-powered showers! It was the first time in several days that many of us felt clean. And the cleanliness had a different feel to us, knowing it had been achieved with clean energy.

– Aftab

Tonight's Digs: Drake Landing solar community, gracious hosts and comfortable beds.

Jodie films Bragg
Creek residents as
they tell their stories

Kealan's birthday!

August 22
Trouble in the Greenhouse

One of the saddest stories I heard on the trip was that of Carmen Ditzler and her failed greenhouse project. As an aspiring agriculturalist myself, it was particularly painful to hear her talk about how futile and uneconomical it is to grow quality food on a small scale. We arrived at her Whiskey Creek Greenhouse in Bragg Creek shortly after dark. Carmen finished putting her kids to sleep, grabbed a leash for her young puppy, then led us into the empty greenhouse, where we sat and listened to her story.

The plants had been removed. The room was dark and empty. Kealan remarked that the space now seemed much bigger than it had when he'd come to buy tomatoes earlier that summer, while working for his uncle up the road in Bragg Creek. Carmen explained to us the process behind growing what she called the "princesses" or

Kealan's family's log cabin

Carmen spoke of regular customers driving down from Calgary in their Mercedes and SUVs, eyeing the modestly higher prices and driving away. She believes that people are simply not prepared to pay the true price for quality food produced ina sustainable manner.

"It's really a lack of vision. I don't understand. Where will we be if we're not growing food? It's a very fundamental thing. Our premier is saying he's not taking his foot off the gas in the oil sands and so that's creating this boom, increasing the cost of everything. Many, many people are being left behind. It's a boom for the few, it's a boom for the government coffers, but you can't see it in the highways, or the schools, or the hospitals. Where is that boom? On farms? Where is the boom? Our neighbours who've farmed forever, who have a small farm, were talking about not taking their hay off this year because of the cost of the fuel to drive around the field. Well, hay is a minimum input crop, other than fuel. So the boom is all well and good, but I don't understand why there's the race to get it out of the ground. It's not going anywhere; no one is going to go along and take it."

"The Americans might," quipped Kealan.

Carmen laughed as she wrestled with the dog. "Yeah, they just might."

– Tim

Tonight's Digs: "The Barn Circle," Kealan's grandmother's log cabin in Bragg Creek

"drama queens" of the vegetable world. A high-yielding yet precarious plant, tomatoes are quite susceptible to disease, especially in artificial conditions such as those in a greenhouse.

You could hear the anger and disenchantment in Carmen's voice, emotions masked only by the love for her family and her desire to create something better. Alberta is awash in oil and money, yet farmers such as Carmen see none of this. As a producer, Carmen took great pride in the quality of her produce. She picked the tomatoes ripe off the vine. Most supermarket tomatoes are picked pink, well before they are ready, so that they can endure the lengthy trip from field to plate. But small-scale producers find it impossible to get their produce on the shelves of the major grocery chains. The grocers sign exclusivity contracts with the large wholesalers, prohibiting them from buying local produce when in season.

Week Two: Calgary to Edmonton

August 23
Our Stay in Sprawling Calgary

I woke up at Kealan's uncle's ranch in Bragg Creek with a knot in my stomach. Our next stop was Calgary, the first city we would be passing through and also my hometown. We had received warm welcomes in the small towns we had passed through thus far on the trip, but I wondered what would happen in this boom town of over a million.

We started the day with a real Alberta ranch tradition: a cattle drive. As we marched the drowsy cows along the little rural roads, Dylan cycled circles at the front, setting the pace and on the lookout for any intrepid runaways. We wondered if it was the first time these cows had been herded by bicycles.

When we set out for Calgary, I urged everyone to pedal hard: we had a date with the media in the early afternoon and about 50 kilometres to go, with a strong headwind and rain conspiring against us. "How strange," I thought. I couldn't ever remember strong winds like that coming from the northeast. The pride of Calgary is our chinook winds, winds from the west that warm as they climb over the neighbouring Rockies and can make even the dead of winter feel like a

Here come the cows!

45

Kalin's rain-shrivelled hands

pleasant spring day. "Oh great," I thought. "Why couldn't we have a warm chinook wind at our backs to whisk us along instead of this cold northeasterly?"

We entered Calgary from Highway 8 and I marvelled how this road has changed over the years. I could remember when turning onto Highway 8 from the city streets felt like turning into the country. Now, the highway was lined with new suburbs and large homes, full of families who wanted their own piece of the prairies. We took note of the names of these new subdivisions: Stone Pine, Elbow Valley, Discovery Ridge, all adorned with "Welcome to . . ." signs built of stone and featuring silhouettes of elk or pine trees. Calgary is approximately the same size as New York, but has only a fraction of the population. With no natural barriers on any side, there is little sign from the municipal government that urban sprawl will slow to a crawl anytime soon.

While this growth continues on the outskirts of the city, the downtown core is largely empty of residents: many office towers but few homes. One Calgarian joked that the only people who live in the core are the homeless. In recent winters, the downtown shelters have been overrun by an ever-growing homeless population that has been left out in the cold by the Alberta boom. With the city rivalling Vancouver and Toronto for the most expensive place to live in Canada, affordable housing advocates have noticed a sharp increase in the percentage of individuals and families who have full-time employment but cannot afford a home — the working poor.

We were met under a rain shelter by the cameras and reporters from several large media outlets. MLA David Swann, the environment critic

Calgary's urban sprawl

for the Alberta government's official opposition, was there to welcome us to the city and lend his support, telling us that our trip was "a strong voice for reason, for responsible development, [and] for the future of the planet." David Swann was mentioned several times by Albertans we met along our way as a strong ally in their struggles for the environment. I was honoured to have his support and explained to my fellow cyclists, many of whom were dubious about having politicians speak at our event, the context of Conservative rule that centrist and left-of-centre politicians, such as Dr. Swann, are up against.

The Progressive Conservatives have held power in Alberta without interruption since 1971. In the 70s and early 80s, under Premier Peter Lougheed, the Conservatives grew in popularity until they held virtually all seats in the legislature. In the 90s, under Premier Ralph

Calgary MLA David Swann greets the cyclists

My frustration with my hometown peaked that night. Besides ourselves and a few close friends and family members, the turnout was miserable at the free event, despite the efforts I knew the organizers had put into publicizing and preparing for it. With not much else planned, we spent the rest of our time in Calgary resting, eating, doing laundry, fixing bikes and speaking with the media. Greg took the prize for "most memorable line fed to the media" when in response to a question on what he expected to see "up there," he responded, "hell on earth."

— Jodie

cial government is gambling: slot machines and VLTs. He said, "It's not that we'll run out of oil, it's that people will have abandoned it as a prerequisite for their day-to-day living, and we'll be stuck with not only all the infrastructure that went into extracting the oil, but the pollution and the tailings ponds and the remnants of the orphan wells."

Not surprisingly, our Calgary stop lacked the intimacy of some of the other smaller locales. In the days leading up to it we'd gained four new riders: Jonathan, Katherine, Lori and Robbie. Still, in a city the size of Calgary, our group felt suddenly small, dispersed, insignificant. Nevertheless, the next day's forecast called for sun, and so I put my best face forward, telling my fellow cyclists not to be discouraged, but rather to see this experience as evidence of the shortage of civic participation symptomatic of Calgarians. MLA Harry Chase echoed my sentiments for Alberta in general, citing the 45 percent voter-participation rate in the previous provincial election, stressing that if Albertans do not participate in the political sphere, they will have to live with decisions dictated to them by unrepresentative governments.

At the crest of a hill on the ride out of Calgary, I turned around and looked out at the skyscrapers and down at row upon row of cookie-cutter houses. We still didn't know where we'd be sleeping that night, but the open road was far more inviting than the claustrophobia-inducing city. I was happy to be back at it. Kalin had driven ahead in the car in search of a place to camp that night. His search landed us in Didsbury, no more than a pit stop on the way to our next big destination: Red Deer.

— Jodie

Our Digs in and out of Calgary: *Historian David Finch's backyard and basement in Calgary; Campsite in Didsbury*

Klein, the party moved further to the right on social issues. King Ralph, as Premier Klein came to be known, was very successful at maintaining the power of the Conservatives until the 2004 provincial election, when the Liberal party gained seats, most notably in Calgary, a traditional Conservative stronghold.

Later that night, we attended the screening of a documentary about urban sprawl organized by some local activists and held at the University of Calgary, where we discussed issues of urban planning with MLA Harry Chase, official opposition critic for Infrastructure and Transportation. Harry Chase spoke of the lack of economic diversification in the province and his fears about the next bust, saying that next to oil and gas, the most important source of revenue for the provin-

August 26
The Petrochemical Plant Near Red Deer

It was raining as we rolled out of Red Deer. This had been a relaxing, home-away-from-home kind of stay for us. We had been taken in by members of the Council of Canadians, offered dryers to dry our wet clothes from the day before, delicious meals, and a lot of food to go. Oh, and beds, actual beds!

It was the first night that we cyclists had spent apart, split into three different groups, and as I huddled on the highway out of Red Deer waiting for the rest of the group to arrive, I felt excited about the return of my new friends, of whom I was completely enamoured by this time. As they trickled in, with rain gear galore, spirits were high.

Two reporters from local television networks tapped their toes and glanced at their watches, waiting for the other cyclists. I held my trusty video camera, filming the TV crews as they filmed the arrival of my friends. The reporters congregated around Tim, asking him, "Why are you riding bikes as opposed to, say, a bus?" Tim explained the symbolism of the bicycle for us: a carbon-free transportation choice that is good for the world and for our bodies. He added, "Let's face it. It's a bit of a gimmick too. You wouldn't have come out here today if we had just passed by in a bus. We are finding that people want to talk to us just because of the approachability of a bunch of young people on bikes."

The Council of Canadians representatives in Red Deer had attempted to get us a tour of the Nova Chemicals Joffre Plant site. But because of unjustified fears of terrorism and environmentalists, we were sent a company-issued promo video instead. We had watched the video the night before and all we learned was that the Joffre plant produces ethylene and polyethylene, the building blocks for the plastics in plastic bags, housewares and milk containers. The location of

On the road to Red Deer

the Joffre plant, just northeast of Red Deer, is strategic because of the large salt mines underneath the site. Ethane, a component of natural gas, and the primary feedstock for ethylene production at Joffre, can be stored in the empty mines during times of low production.

Heading east out of Red Deer, we were told to watch for a turn to the north that would take us directly past the Joffre site. "Don't worry," we were told. "You can't miss it." A tough climb up a curving hill, and the skyline of smokestacks and twisting tubes of the ethereal petrochemical plant came into view. My thoughts were punctuated by the sporadic stream of pickup trucks moving in and out of the plant. The grey cloud growing out of the smokestack seemed to snake its way across the road and hover above a little

Wildlife Photography

During our long hauls through the Albertan countryside, I passed the time keeping a lookout for wildlife with my camera at the ready. Aware that Canada's wildlife population by far outnumbers its human population, I was looking forward to capturing some downright majestic photos. As I soon learned, however, roadside wildlife is smelled long before it is seen, and by the end of our trip I had glimpsed perhaps every species Alberta has to offer at an unbelievably close and stinky range: road kill.

No road trip can be complete without it: the concomitant jokes about who forgot to shower, the inspiration it stirs in any truly resourceful cook, and the musing on the age-old question of "why the chicken crossed the road." And for the less facetious of us, there is the remorse that our comings and goings have such gory side effects. Selfishly, perhaps, what I felt instead was disappointment that I had been cheated out of my chance to see these animals in all their living glory. I felt as if I had set out to see a beauty contest and instead found myself at an open-casket funeral.

That these animals had indeed been glorious was obvious; even in death they retained a vestige of their original beauty. For instance, the deer, from a distance, might as well have succumbed to a fit of the profoundest narcolepsy for all of the serenity of its pose.

Conventional wildlife photography portrays nature as few of us will ever have the privilege to witness it. For all intents and purposes, the image of the bugling elk in the autumn glade might as well be a Playboy centrefold to the average wildlife enthusiast. (Well, okay, maybe not for all intents and purposes.) As a counter to these most unnatural portrayals of nature, let the images presented here be the first in a series of collections that capture wildlife as it really exists in the not-so-wild world that we are creating — dead along the road.

— Greg

50

farmhouse directly across from the site. Also nearby was a small pond, complete with cattails and a few ducks. The air smelled oddly sweet, like roasted pecans.

I was not struck so much by the grandeur of the Joffre plant as by the small house across the road from the facility. I wondered to myself: Which came first, the farm or the plant? How does the family living in that house feel about the plant? Do they worry about air quality? Does the increased pickup truck traffic worry the parents when their children play outside? How can a family be compensated for the changes that the Joffre plant site just outside their front window causes? It wasn't until we reached Fort Saskatchewan, the community that is home to Upgrader Alley, that we would get a chance to ask questions like these, but for now I biked along quacking with the ducks. They, at least, didn't seem to mind.

— Jodie

Tonight's Digs: Homes of Council of Canadians members and friends in Red Deer.

August 27
Landmen, Freeholders and the Fight Over Land Use Rights

Aftab and I had fallen behind, stopping in Ponoka to make use of the Internet services. When we finally arrived at Richard McKelvie's place just outside town, after first taking a wrong turn down a muddy dirt road, we spotted Jackie and Dylan trotting along on the backs of two well-behaved steeds — as though we hadn't been spending enough time on saddles.

Richard McKelvie is a welder, machinist and artist who employs scrap metal to create works that are functional, beautiful, or both. His daughter, Robyn McKelvie, a skilled equestrian,

Jackie is a country girl at heart

was gracious enough to let us ride the horses.

At that moment, Jackie's love affair with the Albertan lifestyle became painfully obvious. Jackie is a city girl by upbringing, but a country girl at heart. She seemed perfectly at ease up on that horse, but was equally comfortable sipping whisky and smoking cigarettes with me by the campfire when the occasion arose.

By evening, over 20 members of the surrounding community, mostly farmers, had gathered at Richard's ostensibly to share a meal of barbecued hot dogs and fresh coleslaw with us, but really, I suspect, to vent their frustrations with the seemingly futile fight over subsurface mineral rights — something most of us cyclists knew practically nothing about.

Richard lives in an underground, hobbit-style dwelling, dug into the side of a hill, just outside Ponoka. That we should find ourselves assembled there to discuss "what lies beneath" seemed more than fitting.

The hobbit house!

Back in Ponoka, Kealan had obtained a phone number for Else Pedersen, President and Director of the Freehold Owners Association (FHOA). I called her up and invited her to join us at Richard's place.

Freeholders are descendants of the original homesteaders who acquired land from the Dominion of Canada, the Canadian Pacific Railway or the Hudson's Bay Company, before these entities began to reserve subsurface mineral resources for themselves. Unlike most landowners in the province, freeholders own the rights to the subsurface minerals, including oil and gas. Some freeholders possess split-title lands and so have access to only selected subsurface minerals, natural gas for example, which was a lesser-prized resource at the time of settlement more than a century ago. The FHOA provides a helpline, seminars and Internet resources to help freeholders protect their valuable, non-renew-able resources. They also conduct research on regulatory and leasing issues and provide a common voice for freeholders in western Canada. Else is a sweet, unimposing woman of perhaps 60 years. I'd heard stories of an elderly woman pushing an Energy and Utilities Board (EUB) official during a heated public hearing. I imagined Else as this woman, sticking it to the man with nothing but her brains and a lot of guts to back her up.

According to Else, the big issue of the moment for people who have clear title to all mines and minerals — apart from coal, or in some cases coal and petroleum — is the ownership of coal-bed methane.

"It's important for you to know what you own before you can even start to negotiate the lease," she told the crowd. "They are trying to attach coal-bed methane to coal, but it's really a natural gas. Just because you don't own the coal doesn't mean you don't own the coal-bed methane," said Else while expressing the FHOA's satisfaction with the EUB's recent ruling that coal-bed methane is in fact gas.

"Do you know anything about subsurface drilling and what do you think about their [the oil companies'] fracking practices?" asked someone in the room, taking the conversation in a new direction.

"We are concerned about those fracking practices. Many of our concerns have been considered out of scope during committee meetings. The only way I could get them on record is by writing them down and submitting them," answered Else.

Fracking, a form of enhanced oil recovery or enhanced coal-bed methane recovery, is the creation of fractures that extend from the well bore into rock or coal formations. These fractures allow the oil or gas to travel more easily from the rock pores, where the oil or gas is trapped, to the production well.

"In a 2,000-foot well they can frack within 10 percent of the surface of the depth of that well, which means they can frack up to 200 feet below the surface, which is in the water area. If they drill through an aquifer and start fracking underneath it, well water tends to run downhill, and if there's a crack, I asked them what would happen and they said they don't care because they know what they're doing," explained one man.

"We were sitting in our house," another began, "and there was a well drilling just beyond our zone and we were having supper and there was an underground explosion that rocked the dishes on our table and our neighbours to the south of the well the same thing. EUB said nothing happened. Apache said nothing happened. Environment said nothing could have happened if EUB said nothing happened. The people who were drilling gave me three scenarios: The rig was up and they weren't drilling; there was too much

The gathering at Linda and Richard's hobbit hole house

snow and they couldn't drill (there was a quarter of an inch of snow); and the rig was down and they weren't fracking or anything. I mean, was I really to believe they were telling me the truth?"

Jodie asked if someone could tell us how the

So what is the average farmer to do in light of all of this? How can he protect his land and himself from the oil companies? How can he hold them accountable for sloppy practices and/or unauthorized techniques? This is what we cyclists wanted to know.

— Tim

Greg's new flute

fracking is done. At this point, pretty much all of the locals were looking to chime in with their two cents' worth. The conversation between them got so lively that I dare say we cyclists could have snuck out without anyone noticing.

"They never use explosives," one man mockingly blurted out, generating quite a chuckle from the room.

"It's pressurized gas and they're basically trying to break up a well that has dropped in production. It may not be explosives, so to speak, but that's more or less the end result," replied someone else.

"Having lived along an ethane line, I've seen them take that gas and shove it down the poor-producing oil wells and the pressure from that pushes the crude to the surface," said yet another.

"They also use acid for anything that has limestone. It burns holes right through and has a tendency to let the gas seep out," we heard from across the circle.

"The biggest thing that the farmer has to do is to get his baseline water well testing done because you're not gonna have a leg to stand on if anything goes wrong and you haven't got that baseline water well testing done. I suggest that they get an independent tester," stressed Else.

"And I pay for it!?" someone retorted.

"Yes, I know, but isn't it worth your peace of mind that you've got a test that you can depend on and it's not industry based?" replied Else.

"The Alberta government has asked all farmers to register their wells and to get a licence for them so that they know what they are producing," added Richard. "If you have that at least you have some history that you can go to. Now how many farmers are taking up that opportunity to license their farms I don't know."

Else's preference is that testing be compulsory for every farmer, for every well. Water well tests can help determine capacity and flow, as well as the presence of gas, heavy metals and other contaminants. But many farmers are reticent about testing their wells, particularly older wells, because the techniques used in testing can damage them and disrupt supply. So, bowing to the farmers, it has been left optional.

Richard suggested that perhaps the discussion was getting off track. Else agreed, but not before adding that "these are part of the logistics that both freehold owners and surface owners must consider in negotiating proper contracts with the industry." As the old saying goes, "the devil is in the detail." That's where they get you.

The whole conversation was indeed a bit confusing for us novices in the room. The oil companies have professional geologists, water experts and lawyers on whom they can call for help, but how does a lone farmer wrap his head around all of this?

"Well, you do this. You sit around and have coffee with your neighbour and bitch and you find out how much he got for his land. That's exactly it. That's the only way we learn really. I mean, there's no training for us. And the oil companies like it that way," someone answered, to the crowd's amusement.

"I think the whole notion of co-operation historically was important and when I look at it now, I think we've somehow shifted in society, that co-operation is not seen as a good thing, it's very private, stand-on-your-own-two-feet sort of thing," expounded Richard.

Richard spoke with more composure than the others, and his comments were broader in scope, almost philosophical. We joked later that we should hire Richard to narrate Jodie's film. He has that James-Earl-Jones type of voice. A large, bearded man, Richard easily commanded our attention as he weaved words to tell the story of the landowners. He was the image of the storyteller I'd conjured up while first conceiving of the idea for the trip months before.

"Now, quite a few people come here and they buy pipe and we do machine work and that sort of thing, and so you do get to talk to people. Murray, the guy who supplies our gas, told me one day of an old Ukrainian farmer. His language skills weren't that good. He was of another generation and he was a trusting sort of a guy. Well, the oil companies came to him and they drilled a well and they found oil and they gave him $50 a year. This had been going on for 40 years. Now, Murray was in his yard one day and the oil man came and gave the farmer his $50 cheque. Murray asked him what it was all about and the farmer told him and Murray said, 'You have oil and they're giving you $50 a year!' Well, Murray just about hit the roof and said, 'You've got to look into this.' The net result was the government forced the oil company to pay him several million dollars. That's how far back it went."

Richard went on to tell another story, but his point was that the trusting person is fair game in the dog-eat-dog world of property rights. It truly is about "what lies beneath" and it's a dirty business, in more ways than one.

– Tim

Tonight's Digs: The amazing "hobbit house," home of Richard and Linda in Ponoka.

Mixed messages...

Our ranks swelled in Edmonton as more bikers joined the group

August 28
Playing to the Media in Edmonton

I remember Edmonton as a whirlwind of a stop. The cell phone rang non-stop that day as we raced alongside busy highways and meandered through the spaghetti streets of suburbia, determinedly making our way into the provincial capital. New riders were looking to meet up with us (our numbers would swell to a high of 19), journalists were seeking to schedule interviews and our interview subjects were trying to confirm appointments for the following day.

Things felt scattered. Cumulative fatigue had made meetings hard to organize and communication was suffering. In Calgary, we had relied on Jodie to guide us from the outskirts, through the mushrooming residential neighbourhoods and the downtown's concrete jungle, to our day's final destination. In Edmonton, we were more or less on our own.

Jodie, Aftab and I arrived at our host Charles's house after a perilous detour down Wayne Gretzky Boulevard. We were expected at my friend Matt's place in a matter of minutes and so no sooner had we arrived than we were being whisked into cabs headed to his apartment somewhere off of White Avenue. Aftab seemed

Jamming at Matt's and square dancing!

content to hit the town in her bright yellow Lycra cycling shirt. I, on the other hand, pulled out my one nice dress shirt from the bottom of my panniers and hastily threw it on, in a ridiculous attempt to compensate for the bike grease on my hands and smell of stale sweat on my body. Aftab has a loving boyfriend. I was single. You do what you got to do.

After dinner at Matt's, we crossed over to a neighbourhood pub, where we met up with the local activist crew for a night of beer and dancing. Dylan grabbed a hand drum and joined in on the Celtic jam. Others grabbed partners and danced their best square dance. I grabbed a beer at the bar.

My faith in the group was restored the following morning when I turned on the television and watched as Robbie, Jodie, Dylan and Danny conveyed our message to the nine-to-five working crowd during a live appearance on *Breakfast Television*. Those who know me best know that I am not a morning person. I brush my teeth at work to save precious minutes of sleep time. I've embraced bed-head, not as a point of style, but as another lazy way to gain some added shut-eye. That anyone would be watching television at such an early hour was unbelievable to me, but even more unbelievable was that my friends had somehow managed to pull their sorry asses out of bed after a night of drinking and dancing and make

August 29
Disposable Workers and Labour Rights

Admittedly, our morning in Edmonton could have gone smoother, hangovers aside. Our meeting with the Parkland Institute was unexpectedly cancelled at the last minute: we were pre-empted by that morning's announcement of plans for a nuclear plant in Peace River. A separate meeting with water expert David Schindler also failed to materialize. Despite the suddenly lightened schedule, we still managed to be late for our meeting with the folks at the Alberta Federation of Labour (AFL). We'd gained five new riders in Edmonton: Jeh, Maya, Danny and the two Joannas (Dafoe and Bruijns). Unfortunately, we'd failed to give them the warm welcome and briefing we like to give newcomers, and the confusion that ensued following the press event at the legislature was the result. Lori had disappeared with the car keys, and my laptop and notebook, (in which I had noted the details of our meeting with the AFL), were in the locked car. When Lori finally reappeared, I ran to the car to retrieve my things, did a quick shout-out for those interested in attending the meeting to pony up, and led the way to the AFL offices. We'd meet up with everyone else later that day in Fort Saskatchewan.

"So you believe in the free market?" Aftab asked Jason Foster, Director of Policy Analysis for the Alberta Federation of Labour (AFL).

"Well, it's not so much that I believe in the free market, it's that I demand logical consistency out of those who demand the free market," he answered.

their way to City TV's downtown studio in time for the early morning segment. Make-up please!!

I myself hurried to print off the press statement and grumpily readied the group for a full day of events, starting with a 10 a.m. press event at the legislature, followed by a slew of interviews with community organizers. Did I mention I'm not a morning person?

– *Tim*

Tonight's Digs: Backyard in suburban Edmonton, courtesy of a Sierra Club supporter.

Press conference in front of the Alberta Legislature

According to Jason, wages in the Alberta oil industry have not followed the basic laws of supply and demand: companies have been able to use a number of tactics to prevent the rise of wages. One such tactic, the Temporary Foreign Worker Program, is of special concern to the AFL.

Although he acknowledges the existence of a labour crunch in places such as Calgary and Fort McMurray, Foster remains critical of both the Alberta government and the oil companies, citing their inconsistencies in dealing with the problem. The AFL has gone so far as to accuse the government of causing the current shortages by refusing to pace development in the tar sands. The glut of new construction, they claim, has led to the current scarcity of skilled tradespeople and the subsequent push to hire foreign workers.

"There are presently more TFWs [temporary foreign workers] entering the province each year than there are permanent immigrants," said Jason. "The entire strategy of the government has shifted away from bringing people to Alberta to allow them to have the full rights of citizenship and become members of our communities.

"They've now shifted it to say we want a revolving door of cattle to do a bunch of work and ship them back home again. They [the oil companies] have found that if you increase supply by bringing in a pool of workers from outside the country who are prepared to work for less and without benefits, you artificially suppress wages."

The numbers seem to support Jason's claims. According to Murray Gross, a spokesperson for Human Resources and Skills Development

Canada (HRSDC), in 2006 Citizenship and Immigration Canada issued a total of 15,172 new temporary work permits for Alberta, bringing the total number of temporary foreign workers in the province to 22,392. In 2005, by comparison, there were 15,815 TFWs working in Alberta.

Jason went on to explain that temporary foreign workers are basically indentured servants.

The permit that allows foreigners to work in Canada has their employer's name on it. Although they are theoretically entitled to the same employment and labour rights as Canadian workers, they don't have the same freedom to act on those rights, since they can be sent home at any time, without question, at the discretion of the employer.

In response, the AFL has hired a lawyer to act as TFW advocate, taking on cases for the workers to help them get their rights.

"We need them desperately, but once they come here, they have no rights," explained Yessy Byl, TFW advocate, during a phone interview I arranged upon return from the trip. TFWs usually get here by dealing with a broker in their home country. The broker offers promises of a job or even immigration status in exchange for a brokerage fee (reportedly between $500 and $5,000). This practice is illegal in Alberta, but it is difficult to stop, since the brokers are generally operating elsewhere. "The brokers charge outrageous fees . . . and mislead people as to what's covered [by the fee] and what isn't. They understate the cost of living. They bring them to Canada and dump them. Often the job doesn't even exist any more," added Yessy.

Temporary foreign workers are guaranteed a set number of work days as stipulated under the terms of their work permits. If a worker loses his job, or leaves it voluntarily, he may choose to seek employment elsewhere. But according to Byl, this is not always as straightforward as it seems.

"If you're not working, it's a minimum of four to five months' wait, up to a maximum of eight months, to go through the process of obtaining new paperwork . . . . Without employment insurance, workers can't very well just sit around until they are able to legally work again. Instead, they work underground for less pay, or they go home."

Back in Edmonton, Jason explained to us that when an employer brings in a worker, it's also the employer's responsibility to find the employee housing. He claims that this too is done with minimal concern or respect for the well-being of the workers. He cites an example of 12 Indo-Italians, brought in by a trucking company, who were put up in a three-bedroom bungalow and each charged $500 a month in rent.

"If you are an employer and you can hire a worker where you can get half of the wages back on rent, that's a bonus . . . . They find these ways to nickel-and-dime them. There are guys that come here, work here for six months, then go home without having earned a penny."

I asked Jason whether the TFWs are successful at integrating into their new communities.

"They're not, and that's by the design of the employers," he answered.

Jason explained that the TFWs are picked up by a representative of the employer, driven to a camp exclusively for TFWs, and that's the last anyone sees of them. Immigrant service agencies, he said, are forbidden under their funding arrangements with the government to serve TFWs. So if they come looking for language classes or information on how to set up a bank account, the agencies have to turn them away or risk losing their funding.

I asked him about credentials. He explained that workers have six months to pass an exam, part of which is practical, part written. He said that they usually fail the written part and excel at the practical. Since the TFWs can work during the six months leading up to their test, employers

generally do not concern themselves with helping them pass the test, since they can easily replace them with a new batch of workers.

"They don't want to invest in training; they don't want to increase their wages. It really is a case of disposable workers."

Jason conceded that there is a range in the manner in which companies treat their workers, and that not all companies resort to using TFWs. Still, the idea that labour is viewed as a commodity to be shuffled around like a deck of cards is troubling. Asked whether the labour movement has been gaining or losing ground in the province, Jason offered the following closing remarks:

"I think we have stemmed the bleeding of basic employment standards. For temp workers, I'm not sure we've achieved much for them yet, but the government knows they can't get away with this forever and hopefully we will see new legislation put into place over the coming months protecting the rights of these workers."

At the end of that part of the conversation the question remained as to what workers themselves could do to ensure their rights were respected.

In an interview with Dru Oja Jay, editor of *The Dominion* paper, conducted prior to the trip, Dru mentioned "speculation that there is a campaign, similar to the war resisters, to actively encourage oil sands workers to slow down or stop the machine that they're a part of."

Although I did not succeed in confirming the speculations, they inspired me: if we really want to fuel change in the tar sands, we can. Workers can demand better, and are currently doing so through the unions. Consumers too must demand better.

There is strength in unity and that's why some companies are seeking to avoid the unions by

There's a shortage of housing...

hiring temporary foreign workers or by using creative, at times underhanded, bargaining tactics. I asked Jason Foster about negotiation and bargaining in the trades sector:

"The world of construction labour relations is Byzantine," he answered. "Even those of us who do this for a living have a hard time getting our heads around it.

"Traditionally, they've sort of had master agreements with specific elements for each of the trades. What happens is that the building trades all sit down around the same bargaining table, but this time around that broke down. So as a result there are a handful of trade unions who have strike votes coming. They could theoretically go on strike within 72 hours. They're choosing to use it as leverage right now, but it would be fascinating if they actually did go on strike. You would see construction stop immediately.

"The basic reality is if any one union goes on strike on a construction site, work shuts down, because no one will cross the picket line. Right now there are high stakes going on."

So could the unions' sway at the bargaining table also be used to push industry to clean up its act? There was a time when the protection of jobs

The wages are undeniably higher in the tar sands than anywhere else. I wonder how other industries are coping with such a gross inflation of wages. You can pay higher wages and offer better benefits, but if you're a mom-and-pop burger shop, you can't very well start asking $10 for a burger to compensate — or can you? People will pay more for oil, but I imagine there is a ceiling for what you can charge for certain things.

— Tim

stood in conflict with protection of our natural environment. Times have changed, or so it seems. The labour movement is leading the charge for a just transition. The unions are now thinking of the long term, using their collective clout to push for sustainable job creation.

"You first need a plan," explained Jason. "The development of alternative energy sources will create jobs as well, in many ways sometimes more sustainable jobs, probably longer-lasting jobs. There are 2,400 workers at Suncor, but when they built Suncor it was probably five times that. Many of the workers in the tar sands are hired during the short-lived construction phase."

While it is definitely gaining traction, realizing the idea of a slowdown in the oil sands remains an uphill battle. The threat of job loss is just too great.

"So why are they rushing so much?" asked Aftab.

Aftab has a way of asking simple questions that generate rich answers.

"The greed of profit and money don't allow for people to wait," answered Keith from the CEP (Communications, Energy Paperworkers Union of Canada).

"These are really capital-intensive projects; the sooner you pay off your capital costs, the faster you see a return on your investments," added Jason.

"In other industries, like auto or mining, there are up and downs; the company that can do the best job survives. The oil sands have never known a shutdown, a slowdown, or a layoff of workers because we couldn't sell our product. As long as people want to buy oil they'll produce it," continued Keith.

"The basic laws of economics do not apply to the tar sands," quipped Jason.

I asked Jason whether he's noticed slowdowns or shutdowns in other industries due to the spike in wages.

"The public sector is really struggling right now because unlike the corporations, they don't have the profit margins to play with, unless you consider surplus a profit margin. So their wages are a lot more inelastic; they have a harder time giving wage increases because it's tax revenue. The public sector is quickly losing people to the private sector. And so in any sector where your prices are more inelastic, you're going have that trouble.

"Nursing, for example, is having trouble attracting people into the profession, due to the nature of the profession, the stress. It's still a predominantly female occupation and often nurses find that they don't need to work when their husbands are making more than enough money in the tar sands."

"There are stories of teachers in Fort McMurray leaving the classroom to work in the mines," added Keith.

"That is so disturbing when you think about it. We've created an economy that has given people the incentive to leave important professions like teaching and nursing to go and participate in the gold rush," concluded Jason.

Workers have a great deal of influence on the future direction of the oil sands. They will play a central role in shaping the industry. That summer, we stood (or biked) in solidarity with them.

— Tim

Week Three: Fort Saskatchewan to Fort McKay

August 29
Passing Out in Upgrader Alley

When the sirens go off at Shell's upgrader near Fort Saskatchewan, Alberta, nearby resident Kathy Radke knows there has been another accident. As plumes of toxic vapour are picked up and scattered by the wind, she is expected to call an emergency hotline set up for the handful of families living in the immediate vicinity to find out how severe the accident is, and whether to "shelter in place" or evacuate the area.

"Half the time, the info hasn't even been updated when we call," says Kathy. She wonders why Shell doesn't supply the nearby residents with air packs that they can put on as soon as the alarms go off.

Since Shell built its first upgrader in 2003, accidents have occurred at the rate of about four or five a year, says Kathy. Last September, there were two gas leaks in the space of one week. Nearby residents were instructed to stay in their homes for several hours. Some later reported sore throats and headaches that lasted for days.

Welcome to Alberta's Industrial Heartland, a 78,550-acre area about a half-hour's drive northeast of Edmonton. This industrial zone, home to dozens of refineries, petrochemical plants and other industrial facilities, is where much of the bitumen pulled from the tar sands is pumped, through a 493-km pipeline, to the Shell upgrader. Upgrading is the process by which bitumen, a thick, tarry muck, is turned into a synthetic crude oil that can be sent to refineries. This extra step is part of what makes the production of oil from tar sands so energy-intensive, with greenhouse gas emissions three times as high as those associated with conventional oil production. More natural gas is eaten up by bitumen upgrading than by the mining process itself. With two more upgraders under construction, and another 10 in

On our way to Fort Saskatchewan

Group shot!

various stages of development, the area is popularly known as "Upgrader Alley."

A few decades ago the area was mostly farmland. And several families still live here, scattered on patches of land between the massive industrial facilities. Our first order of business in Fort Saskatchewan was to attend a community barbecue in the park, hosted by the local chapter of the Council of Canadians. It was there that we met Kathy Radke.

The Radkes live and farm on land that was bought from Kathy's in-laws in the 1980s by Atco Gas, for the salt mines underneath in which they store natural gas. When Kathy and her husband moved in, leasing their home from Atco, they were told they would be able to live and farm

there safely for decades. But with the rapid growth of industry in the area, the Radkes were soon surrounded by the clanking of machinery, heavy truck traffic and air pollution. Their house is two kilometres east — and downwind — of Shell's massive Scotford operation, which boasts the only upgrader in operation, a second under construction, and a refinery. To the north, BA Energy is also building a new upgrader.

Shell's neighbours are exposed to routine emissions of sulphur dioxide and other toxic gases, which briefly spike above regulated maximum levels on a regular basis. Over the past two years, Kathy's family has lost 45 dairy cows out of a herd of 140, and she suspects that the air pollution has something to do with it.

The region's flurry of upgrader construction is a direct result of the increase in extraction from the tar sands. Some of the upgrading is done on site in Alberta's north, but with its lower costs and easier access to workers, the industrial region northeast of Edmonton has become the place of choice for upgraders in the province. The bulk of Alberta's bitumen, however, is still upgraded in the United States — where the vast majority is also refined, sold and consumed. Alberta politicians have been calling for a dramatic ramping-up of the province's upgrading capacities. "If we insist on just sending raw product out of this province and adding value to that product in another jurisdiction, the taxes on the value-added product will be paid in that jurisdiction, not in the province of Alberta," Premier Stelmach told reporters last December. Energy minister Mel Knight has been quick to reassure the public that Alberta should have enough capacity to upgrade about 80 percent of its bitumen within a decade or so.

Goofing around before the tour

Our tour of Upgrader Alley

Upgrader Alley by night

Albertans face a dilemma. With the havoc created by the mining and in-situ extraction of bitumen from the tar sands, they already shoulder the brunt of the pollution created by the wresting of oil from tar sands, while capturing only a small fraction of the profits. Greater involvement in the value-added process of upgrading would increase the public's economic return, but it would also concentrate even more of the pollution in Alberta.

There is also the possibility that the upgrader boom could overwhelm Fort Saskatchewan and other towns in the area, in much the same way that Fort McMurray — epicentre of the tar sands extraction operations — has already been overwhelmed by economic growth that most locals say is too much, too fast. The growing pains experienced by Fort McMurray include skyrocketing housing costs, a homelessness crisis and a severe shortage of health care and other services. "Our water treatment plant will be at capacity next year. Our recreational facilities are overtaxed. Our landfill site is full," the city's mayor recently told a parliamentary committee. Is this what Fort Saskatchewan has to look forward to?

"What we're facing is a huge expansion of roads, infrastructure, sewer systems, water systems, bridges. How is that going to be paid for?" asked Edmonton mayor Stephen Mandel in 2006. He estimated that the money invested in the last five years in Fort McMurray pales in comparison to what is proposed for the "Industrial Heartland."

For now, the protests of Kathy Radke and the handful of other residents directly affected by the upgraders are largely overshadowed by the prospect of jobs and money rushing into the region. But if things go the way of Fort McMurray, it won't be long until all of the area's residents experience the ugly side of Alberta's bitumen boom.

– Lori

August 29
Sleeping in Alberta

For me, one of the most memorable things about the To the Tar Sands bike trip was the variety of settings in which we slept over the three weeks. Early in the trip someone introduced the practice of taking naps on the side of the road, whenever one felt like taking a nap. No tents, no sleeping bags, just you in the ditch or on a field, under the big sky. Soon it became a favourite pastime. By the time we were north of Edmonton, the naps on the side of the road became particularly

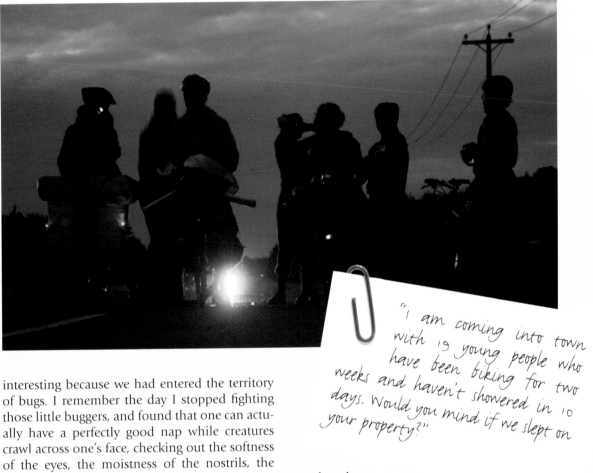

"I am coming into town with 19 young people who have been biking for two weeks and haven't showered in 10 days. Would you mind if we slept on your property?"

interesting because we had entered the territory of bugs. I remember the day I stopped fighting those little buggers, and found that one can actually have a perfectly good nap while creatures crawl across one's face, checking out the softness of the eyes, the moistness of the nostrils, the mystery of the ears. Doesn't the average person swallow hundreds of spiders in their sleep anyway?

Bunking down for the night was a whole different story, and quite an adventure in itself. Our nightly accommodations ranged from queen-sized beds with swan feather pillows to make-shift campsites in the middle of the forest and absolutely everything in between. Shawn, Kealan and Tim had done quite a bit of work before the trip, phoning friends of friends and arranging for accommodations. (I imagine the conversation: "Hi, I am with the Sierra Youth Coalition. I am coming into town with 19 young people who have been biking for two weeks and haven't showered in 10 days. Would you mind if we slept on your property?") Between Waterton and Edmonton arranging for accommodation had been relatively easy because southern Alberta is fairly densely populated and there were both many friends and many campsites to rely on. But by the time we were into the third week of the trip, we had fewer places set up and we were often asking locals for tips on where to sleep the next night.

Possibly the craziest sleeping adventure we had happened just outside Fort Saskatchewan.

After hearing Kathy's harrowing tale and others much like it, we were ready to call it a day, but as usual time had managed to get away from us, the sun was setting and we still had a good 30-kilometre ride to our campsite in Elk Island National Park. Could we not stay somewhere nearby? The locals suggested a campsite up the road. "No more than 20 minutes," they said. We sent the car ahead to investigate. Soon thereafter, the car returned, informing us cyclists that the site was more like an hour away, but that, luckily, they'd found an alternative place to make camp.

The car had stopped at a perfectly nice-looking park. A "Wildlife Greenbelt," according to the signs put up by the sponsoring company, Dow Chemical. Dow Chemical is a manufacturer of petrochemical products (plastics, fertilizers, etc.) with a huge presence in Upgrader Alley and several plants around Fort Saskatchewan. Our hosts at the barbecue had just finished telling us about the effects of the petrochemical industry on the health of the plants, animals and people in the area. We certainly could feel the effects of the air pollution on our own bodies around Fort Saskatchewan; those of us with asthma had to use puffers for the first time in two weeks.

The other story we heard was about Dow Chemical and other petro-related companies' involvement in the communities where they set up shop. It was no secret that most community facilities and services in the smaller cities and towns in Alberta are sponsored by big companies directly or indirectly involved in the oil industry. At some point in the trip, one mother had told us that if it weren't for the support of big oil companies her kids would have no soccer teams to play on. We had benefited from this kind of support ourselves — whether we were conscious of it or not — when we were treated to a free shower at a recreation centre somewhere, courtesy of big oil. That is what Alberta is like. At first glance there appears to be a symbiosis between petroleum companies and citizens. I have since spent a lot of time reflecting on the true nature of this relationship.

At any rate, we had come upon the Dow Chemical Wildlife Greenbelt and we were exhausted. The park was technically "closed" according to the hours posted, but there was no barrier. There was also no evidence of "wildlife," but we figured that the exposure to chemicals could not be any worse here than anywhere else in the area, so we decided to set up camp for the night.

I remember the moon was amazing that night. It was full and orange, large and low. We were surrounded by industrial structures covered in lights, which were beautiful in their own way. Dylan and I hung out on the side of the road with flashlights for a while, waiting for rest of the team to arrive. Some time later, we noticed that Shawn hadn't shown up.

Shawn had his own unique travelling habits. He was usually one of the first to get on the road in the morning. He would ride as much as he could and when he felt "dead," in his words, he would hitch a ride. One time a woman actually pulled up beside him and strongly suggested that she give him a ride without his even asking! As far as I could tell, Shawn never slept in a tent the whole trip. I don't think he even used a sleeping bag. He would simply curl up into a ball, by the fire at a campsite or under a chair in someone's basement, put on the hood of his rain jacket, and fall asleep. But he was always close by. This was the first time he was missing at night.

A search party went out looking for Shawn that night. Jackie and Tim drove up and down the road for over an hour, while the rest of us waited nervously at the campsite, cell phones on hand in case of a call, feeling awful about having lost one of our fellow travellers. As it turned out, we didn't find Shawn that night. I'm not sure if he'd passed by without seeing us, or if he had given up before reaching the group. But he had gotten lost and decided to get into his familiar fetal position in the ditch on the side of the road. I guess for him this was not so different than any other night, but I will never forget how worried we were about him, or how happy we were to see him when he showed up at the Marie Lake protest the next day.

I am particularly sorry that Shawn missed the piece of hilarity that took place that morning before we met up with him in front of Premier Stelmach's constituency office.

At around seven, I woke up in my tent to the sound of tires crunching and what I thought were hundreds of police cars ready to arrest us. I began swearing under my breath. Jodie, who shared my tent, woke up. She whispered, "What's going on? Are we late? Did we miss an interview?"

Our voices drew the officers to our tent. There was a knock on the flimsy tent pole: "Excuse me, sorry to wake you up. Can I speak to you?" I peeked out and there stood a guard employed by Dow Chemical, not a police officer. And he was the nicest, most apologetic enforcer of the law I'd ever encountered!

"Sorry to pick on you, ma'am, it's just that your tent is the first I came across. I am very sorry to inform you that you are actually on private property and you have to leave. This is not a campsite, and I apologize for our poor signage. It was our responsibility to better inform you last night; I apologize for that. Sorry to wake you up so early, but could you please call the rest of your group and take down the tents as soon as possible? Again, I apologize for the inconvenience."

I began to give him the story about how we were lost and tired the night before, but there was no need. He wasn't going to arrest us, or give us a ticket. And he had actually waited until 7 a.m. to wake us up! He and his buddy trusted that I would get the troops moving and immediately left the scene. There were giggles from the

Quickly packing after the guard shows up!

neighbouring tents as soon as the guard left. I was relieved at how pleasant our expulsion had been. "All right guys, we'd better get out of here."

Our standard wake-up call on the trip was Kealan playing his guitar. He is one awesome musician who sings about addiction to oil and rocking in the free world. Sometimes he would be accompanied by the heavenly voice of the beautiful Katherine. The music was one of the best parts of the trip. That morning, Kealan began playing the guitar before he had even put on his clothes. He sat naked in his tent, and played his wake-up songs to us, looking out through a narrow opening to see that we were in fact getting up. We all thought it was hilarious. Laughter first thing in the morning: It was going to be a great day!

– Aftab

Tonight's Digs: Emergency camping at Dow Chemical Wildlife Greenbelt outside of Fort Saskatchewan.

August 30
Solidarity with the Residents of Marie Lake

It had taken Leila from the Sierra Club and me a full day of phone tag to coordinate, as I rode from Red Deer to Ponoka several days earlier. The folks at Marie Lake were planning an action outside Premier Stelmach's office and wanted us to attend.

I'd first heard of Marie Lake a week earlier during our stop in Okotoks. Ted Morton, the local member of the provincial legislature and minister for Sustainable Resource Development, was scheduled to speak at a town hall-style event in Turner Valley, and so Jodie and I had decided to drive back to Turner Valley that evening from Okotoks and catch it on film.

Lindsay called my cell just as we were leaving. I asked her if she had any questions for Ted Morton.

"Ask him about Marie Lake," she said. "He just approved seismic testing up there and the cottage owners are pissed."

"Murray Lake? Okay, got it. I'll ask about that," I obligingly said before hanging up.

And so we went and I asked about the seismic testing. Luckily, people there knew less about the northern lake than I did and no one picked up on my obvious ignorance of the issue. Thank God for homonyms!

We learned from Morton that an oil company had bought the subsurface mineral rights under the lake. Whereas the law prohibits drilling in the actual lake, the company was proposing to start about half a mile back and tunnel underneath it, using steam-assisted gravity drainage (SAGD) to get the heavy oil out. But before that development could proceed, seismic testing was required to verify the location of the oil deposits. Morton's department is not responsible for regulating drilling; however, it is responsible for seismic testing. According to Morton, there were two issues: Does seismic activity on Marie Lake threaten either the lake or the wildlife in the lake? Second, should the development take place?

"You have to let the policy decisions follow a process. If you just have one-off decisions made by this minister or that minister, because they get too enthusiastic or scared by something, you'll have policy chaos. There's a process in place for a reason, to give predictability not just to the oil companies, but to the citizens of Alberta," he said in response to my question about the government's decision to allow seismic testing on the lake.

The Stelmach government's new Land-Use Framework was designed to provide an approach to better manage public and private lands and natural resources to meet Alberta's long-term economic, social and environmental goals. How well it succeeds in balancing the competing interests of the many stakeholders involved is yet to be seen.

"I'm not for or against the Marie Lake development. What I am for is allowing the seismic to take place unless there is any evidence against it. I sympathize with the cottage owners there. If I was a cottage owner, I'm sure I'd be trying to stop the whole thing too," Morton continued.

Morton is well known as an outdoorsman. It was interesting to watch him talk his way out of some of the other questions posed to him that evening, most of which revolved around the recreational use of all-terrain vehicles on Crown land. It's interesting to see who takes advantage of forums such as this one. The four-wheeler crowd definitely had its voice heard on this particular night.

Back in Fort Saskatchewan, after being awakened by bemused security guards at the Dow Chemical Wildlife Greenbelt, we politely left the premises and made our way to Premier Stelmach's constituency office downtown. The cottage owners, consisting of maybe half a dozen families, were already gathered there, carrying cardboard cutouts of fish with slogans like "Save Our Lake." The media were out in full force, at least a dozen reporters. We recognized some of the journalists from the previous day's press event in Edmonton. A bunch of us chanted with the cottage owners and joined them inside as they spoke with the secretary, demanding a meeting with the premier, who of course was out of the office.

Some of us spoke to journalists; some of us

Marie Lake protest

The long road north

played the role of journalist ourselves; some of us sat in the shade eating breakfast.

Several days later we received an email from the protest's main organizer thanking us for our support on that day. The decision had been overturned: there would be no seismic testing on Marie Lake.

There is a certain not-in-my-backyard quality to the Marie Lake example. The success of the uproar of a handful of cottage owners in stopping the testing will have set a precedent for similar cases. It is one instance where our presence and numbers had a clear impact, but more importantly, it raised questions about how and when we should work with people actively engaged in their own communities to effect change.

The question of solidarity work — when to lend our support to other grassroots causes or campaigns — was central to the advocacy-versus-storytelling debate that took place among the organizers of the trip. It's hard to think of an issue as complex and multifaceted as the Alberta tar sands. Would it not have been a bit naive to parachute in and take position on it when we had barely even begun to try to understand it? We wanted to know what we were dealing with first. We wanted to learn from the communities living with the impacts of tar sands development. We wanted to learn how to engage in their struggle in a way that is respectful, takes direction from the people on the ground and reinforces the work already being done.

Still, the nomadic nature of the bike trip made it difficult to really get to know our allies. It takes time to build relationships of trust and solidarity. This may explain why some of us munched on fresh fruit in the shade, while others marched into the premier's office. It's easy to feel disengaged when you lack a personal connection to the situation at hand. Perhaps that's why the majority of Canadians can continue to list action on climate change as important, but fail to make it a priority in their daily lives. They've yet to truly feel its impacts.

One way we can continue to work in solidarity with Albertans living the down side of tar sands development is to spread the word and inform the rest of Canada about what is really going on there. Information is power. God knows the big oil companies hate the attention. Why else would they ban cameras on site or spew the most toxic of their waste at night?

— *Tim*

Tonight's Digs: Elk Island Retreat, inside the biggest tipi we had ever seen.

Camping at Lac La Biche

August 31 to September 4
The Long Road North to Fort McMurray

Heading slightly east and then north again from Fort Saskatchewan, we'd made our way to breathtaking Lac La Biche, where we were treated to a beautiful peach sunset and a night of saunas and deep conversation at the Skyke family farm, a stay arranged for us by Lindsay. The next morning, several of us surveyed the map thinking ahead to the next few days of our journey. Civilization was a thing of the past, we joked. And now our trip was to become forest and highway and little else.

"What's this?" Maya asked, pointing to the dotted line that started a third of the way to our destination.

"That's the legend for gravel," Katherine explained.

"Gravel? The road isn't paved?"

"We were told," Kealan mentioned as he strolled by the assembled group, "that it's been paved since these maps were made."

"Let's hope so," Maya said as she and I exchanged concerned looks. The question in the back of our minds: would we make it, riding on a gravel road that ran for 100 — maybe 150 — kilometres? Perhaps we were fretting too much, but at that point in the trip it felt like it might break us.

Our final route had been the subject of much discussion before our trip to the tar sands began. At the time we'd only known of one path to take between Edmonton and Fort McMurray: the

A curious trucker stops us

dreaded Highway 63, infamous for its danger to motorists, unthinkable for cyclists. Highway 63 is not patrolled, we were told. We did not know whether to believe this particular story; we hoped the police had not given up entirely. Memorial crosses, we were told, lined the highway as remembrances of the fatal accidents that had taken place there. Even in automobiles the roadway is to be avoided at all costs on Thursdays, the first day of the tar-sand worker's weekend. It was said that they race down to Edmonton to spend every bit of their free time in "the human world," drunk or sober. Why? Only those who had "lived" in

the camps of Suncor and Syncrude, "hell on earth," understood.

But we were not going to find out if the legends were true. We put the map away as we prepared to leave Lac La Biche. From here, we could truly only take one direction: north to Fort McMurray, the long way, the safer way, on Secondary Highway 881.

The late afternoon sun had fallen behind the trees while our pedals swooshed continuously. We'd left Lac La Biche late because we'd become embroiled in an argument about what we hoped to accomplish by rolling into Fort McMurray. This was our collective existential crisis. What exactly was the point? There were those of us who felt we should arrive in Fort McMurray with a definite position, a sort of official proclamation stating that after three weeks of cycling we were ready to assert that the tar sands development is bad and needs to be stopped. Others felt that, if anything, the trip had raised more questions than it had answered. Calling for the shutting down of the tar sands, as Alberta newcomer Greenpeace had recently done and many of us had previously endorsed, now seemed too simplistic. The forces at play were just too powerful for us to imagine an easy resolution to the mess we were in. The tar sands, we had come to realize, were part of a much larger global machine. It was hard for many of us to imagine a domestic government with sufficient gall and resolve to shut down the tar sands and save the planet at the expense of the global economy. If you consider that Canada's political stability and friendly business climate are the chief reasons why tar sands oil is so attractive (that is, countries don't need to fight wars for it), then foreign military intervention becomes the only imaginable response to a complete shutdown of the tar sands. Still, a complete shutdown down of the tar sands is unquestionably what our planet needs in this time of climate crisis. The feasibility of a

ENTERING

HEART LAKE FIRST NATION
TRADITIONAL LANDS

PLEASE HELP US PROTECT THE ENVIRONMENT

74

The last stove for 100km

shutdown in practical terms is a whole other question. In the end, we would enter Fort McMurray without a clear and common position on the tar sands, saying only that the status quo was unacceptable and that we'd witnessed the existence of a silent majority unhappy with the current pace of development and itching to to be heard.

And so, after hours of debate, we'd left late, the sun high in the afternoon sky. The forest floated by, as we circled the sparkling lake that marked the southernmost point of 881, and rolled onto the highway. Our countdown (250, 249) was under way.

Not long into our ride that day, we passed the banner we'd used so often to declare our meeting site. I pulled to the side of the road and hopped off my bike. I ran back to meet Dylan, who was already walking his bike across the ditch and into the great Northern Boreal. There, in the woods, I saw the car parked and the tents of the first cyclists, arranged sporadically.

"Why so soon?" I called to Katherine, who had gone ahead in the car to find us a place to sleep. She explained, "There's a gas station at the intersection ahead, and after that there's absolutely nothing for the next 100 kilometres. I thought it safest to stay here and do them all tomorrow."

In the southern prairies, we had passed through vast fields of ranch land and country-side, land that knitted into the fabric of human society. The next morning as I rode onwards, I felt as if I was at the edge, the frontier of my world. I felt in those moments that I was seeing what the first Europeans saw, when we arrived nearly 400 years ago and mistook this place for an empty land because it was not scarred by human ownership, separated into "humanity" and "everything else." This may be the world we're trying to save, I thought.

"You know, the magic of this place is lost for me when I remember what they told us," Aftab sighed. We sat along the highway, looking out into the infinity of the boreal forest. We spoke softly, to each other and to the forest as well.

"What do you mean?" I asked.

"They fake it, the companies. They know that the people who pass through here love the forest, so they don't clear-cut near the highways. They leave the forest for a mile or so; that way it looks pristine to passersby. Why protest if everything looks healthy? There's not as much forest back there as we think."

Word of mouth spread the events of the trip slowly among the bikers. At rest stops, through cell phone conversations or at the end of the day, we would share what we learned each day. On our second day on 881, we began to pass SAGD oil mining sites and logging company outposts. We didn't know much about SAGD yet, but we did know this wasn't the oil drilling we knew from pictures and stories.

The day after we left Christina Lake

The wilderness. The great Canadian North. It felt like the first time I had seen my own country, the Canada of our collective dreams. A vast emptiness, owned by no one.

– Kalin

Travelling on a bike is different from other ways of moving from one place to another. Different from cars, even from trains. On a bike you are in, you are part of. No longer an observer from a world removed, you are present. This can be wonderful, but also frustrating and sometimes dangerous.

— Kalin

a new story rolled through the group of cyclists: a worker from a nearby work site paid us a visit at our Gregoire Lake camp-site.

"There is a security memo circulating at the site that the cyclists are not to be allowed on the premises, that they should not be per-mitted to talk to anyone," our guest told us. His story ignited feelings of frustration among us. A few riders broke into tears, brought on partly by fatigue, partly by the stress of the situation. We were terrorists, according to the memo, hiding our true intention to destroy industrial property.

The notion was hilarious in a way, that we could have ridden for a thousand and a half kilo-metres, engaging in public discourse, pondering the problems of the oil economy, whilst hiding Molotov cocktails and homemade bombs in our saddlebags. It was absurd.

We tried to attribute it to miscommunication. We had stated our message time and again in the media. We had actively tried to separate our Sierra Club affiliation from our goal of getting at the truth. What we were loath to admit to our-selves was that the oil companies in Fort McMurray clearly saw the matter differently. At no point up to then had they officially shown any interest in talking with us. The label "terror-ist" was either a useful tool to frighten security teams, or an accurate measure of how dangerous we truly were to them: to even consider the ques-tions we were asking, to slow down and examine the how at the expense of the for what would

truly damage them. They saw us for what we were; they knew we were not there to help them continue doing what they wanted to do in the way they wanted to do it.

On breezy, sunny afternoons we felt free and happy, our moods reflecting the atmosphere of the earth, glad for the gentle whispers of support it gave us. On rainy days we felt suffocated, pushed towards the ground by the weight of cloud and water. The wind took on an impor-tance in our lives it had never had before: a headwind was an invisible barrier that held us back from our daily, hourly goals. Hills and mountains were no longer simply natural vistas, but direct arbiters of energy use and exhaustion. It was hard at times to feel affection for the plan-et that dropped these obstacles in our path. Part of the wonderful and the miserable. Belonging to the world. In truth, I have rarely felt as alive as I did during my time on the road from Lac La Biche to Gregoire Lake.

We had furiously debated the pros and cons of finishing 881 in a massive burst of effort; in the end we decided to cycle the second half of the highway in one day, covering the territory from Conklin to Anzac. It was to be our longest day (125 km), and when we awoke that morning to the rapid patter of raindrops on tent flaps, we realized it would be the hardest as well. We set out nonetheless, conscious that if we could fin-ish this one, obstinate leg of the journey we would have made it.

It was the day my bike decided to break down.

Half an hour into the ride, swamped with wet grit from the highway, my rear gear-cogs began to grind, and as I confusedly pressed the shifter, I tried to figure out what was wrong. Suddenly something fell from my bike, and those sur-rounding me stopped while I tried to find the problem. Then my stomach sank: the shifter itself had broken off, irreparably. I could no longer change gears, and there were over a 100 kilometres left to conquer.

My first thought was how screwed I was. My trip was over, this bike was essentially unrideable, and I wouldn't make it. The others went on ahead while I waited with Kealan for the support vehicle, and tried to fix the bike. The best we could do was to set a gear manually — by moving the chain on the cogs by hand — and stick to it. I nearly gave up, but when the car rolled around, I realized I couldn't imagine finishing off the 1,500-kilometre journey in the passenger seat of a car. I dazedly refused the lift, and got back onto my bike, aware of the difficult ride ahead of me.

That day proved, miraculously, to combine all of the barriers we had faced up to that point. The pouring rain eventually ceded to sunny sky, but fierce wind sprang up in its place, and every so often a cloud of rain would roll over our path again, soaking us briefly before continuing on. At the lunch break, Jeh photographed the beard of mud I had accumulated; protected by sunglasses and helmet only the lower part of my face had been completely covered. The others laughed at my new facial hair, which helped me relax. Otherwise, my day was only pain.

Eventually we came upon the rolling hills of the Athabasca region, and coupled with the wind and my rigid gear, I became the turtle in the race. I grew frustrated with myself and the others; my mood was like an emotional whip, lashing my body into forced labour.

It became apparent that we had underestimated the distance to Anzac; by this time alone except for Jonathan's persistent and saintly company, I nearly gave up on my quest to finish the trip on my bike. But somehow, as I continued hour by hour, expecting each time I sagged into a five-minute break that I wouldn't be able to get up again, I reached into pockets of energy and determination that I hadn't known I had.

Kalin's mud beard

As the sun went down, 12 hours after we'd left Conklin, we rode into another rainstorm — and our destination. I pitched my tent in silence — furious at nothing in particular, wild with exhaustion — managed to eat dinner, and collapsed. But at least I made it all the way, I thought to myself. Tomorrow I will be okay, tomorrow I will arrive in Fort McMurray. We made it. We're here.

– Kalin

The Week's Digs:

August 31: Cabin in Long Lake Park, courtesy of a generous mother and her energetic eight-year-old, Ty.

September 1: Camping on the Skyke family farm, featuring a surprise sauna.

September 2: An opening in the boreal forest on the side of the highway.

September 3: Christina Lake campsite in Conklin, with remarkably nice showers, bathtubs and laundry facilities.

September 4: Gregoire Lake campsite in Anzac, struggling to make a fire in the rain.

September 5

The Good, the Bad and the Ugly of Fort McMurray

The Bad. Fort Muck — place of sex, drugs, violence, homelessness, massive trucks, polluted air and contaminated water. This is what we were told we would find at the end of deadly Highway 63, or in our case Secondary Highway 881. The city of 70,000 has been growing at a most alarming rate. In response, City Council has gone so far as to call for a moratorium on new developments. The municipality simply cannot keep up with the endless stream of new arrivals and the associated demand for services.

Back in Fort Saskatchewan, I'd asked our contact Lyn Gorman about the rumoured culture of sex and drugs up in Fort Muck. Before moving to Fort Saskatchewan with her husband three years ago, Lyn spent 27 years working as the executive director of the Wood Buffalo HIV & AIDS Society. Her husband Keith worked for Suncor and now works as a rep for the CEP union there. We'd met him, along with some other knowledgeable labour folks, back in Edmonton. Lyn currently works as the Prairies Coordinator for the Council of Canadians. Needless to say, she is well acquainted with the goings-on in Fort McMurray.

"The drug of choice changed overnight, from pot to coke, and it changed with good reason," began Lyn.

She explained that cocaine, being water soluble, leaves the system within 24 hours, whereas pot is fat soluble and so traces remain for weeks.

When companies began instituting mandatory drug tests, the dealers starting pushing coke instead of marijuana. I asked Lyn for her take on the high levels of drug use in the relatively small community:

"They have the cash. Blue collar, white collar, ordinary people — it's everyone really. There are parties where you go upstairs to do heavy drugs and downstairs to drink."

The sex trade in Fort McMurray has kept pace with the booming oil industry, and goes hand in hand with the increase in hard drug use. Escort services are the norm, as is evident from the array of advertisements in the phone book — "from two to 12 in only a matter of years," Lyn claimed. "Strolls" are less common, but can still be found. There is concern that the high-risk sexual behaviour is fuelling an epidemic of disease, notably syphilis and HIV. Sexual health statistics for Fort McMurray are near average, but, according to Lyn, most people seek treatment in Edmonton or Calgary, so it is impossible to accurately track transmission.

"I've had boys barely out of high school come into my office freaked out over their sexual identities after getting drunk and having gay sex for the first time, and often unprotected. It's just plain sad to see."

"Does everyone do it [engage in high-risk sexual behaviour and drug use]? No, but the fact is we have a syphilis outbreak in this province, and these workers go back to where they are from and it spreads. There are health centres at the work sites, but people don't go to them for fear that it'll get back to their employers. It's an unacceptable situation, but the municipality has bigger things to think about and so the problem is not dealt with effectively."

During our time in Fort McMurray, we stayed with Morgan, her husband Bruce, and their two young boys Marshall and Kane. Morgan recently left her job as a tour guide at the Syncrude facility to stay at home with the kids. Bruce operates heavy machinery at the same Syncrude facility. The boys like to watch *Shrek* and wield light sabres.

While genuinely concerned about the future of their children and the environmental impacts of the industry they work for, both Morgan and Bruce demonstrate a great deal of pride and passion for their community and its backbone, the oil sands industry.

"It's human nature to focus on the negative. It makes for a better story. I just wish people would realize that there is an actual community here in Fort McMurray with good things happening. People come here and they are surprised to see that we have a hospital and a

The Good. Surely there is some truth to the horror stories floating around, but, as we discovered during our visit to the booming oil town, there are two sides to this, as to every, coin.

Trick riding into town

The Ugly. The ugly side is its seemingly unshakeable reputation for destruction and greed. There is talk about upping production in the tar sands to five million barrels per day by the year 2030. True McMurrians might prefer a slower-paced development for, unlike the shadow population, they aren't going anywhere.

library and a YMCA. It's like all they expect are a few rundown shacks, a bar and general store," Morgan said.

"You can find sex and drugs in every community, if you look for it. I don't worry about that so much. My main struggle is to teach my kids the value of a dollar. There is just so much money here. When the tooth fairy is giving out $10 for a tooth, how do you explain to your child that his tooth is worth only a buck?

"Sure, we have our problems, but they are no worse than anywhere else," continued Morgan. "Here in McMurray, we have what's called the shadow population — people living here for six- to 12-month stints, or commuting from away. They don't pay property taxes, yet they use all the services, the roads, the hospitals, the water, the swimming pools, et cetera. They aren't invested in the community, they don't respect themselves and they don't respect their surroundings. They give the place a bad name. Meanwhile, there are people like us who feel a real sense of pride in the community here. We are invested and this is our home, but no one asks us what it's like here."

Lyn's and Morgan's views of Fort McMurray are so radically different that it's hard to believe they are talking about the same place. The truth is McMurray is a city caught in the middle — its fate determined by its geology. It just happens to sit on one of the richest oil deposits in the world at a time when global demand for the stuff has

never been higher. It did not choose its destiny, but it has done its best to assume it.

The municipality has been forced by the rampant growth to react to issues as they pop up instead of pre-planning, but even that is changing now. Beth Sander, manager of the municipal Planning & Development Department, was proud to tell us that "they are planning to look ahead, to work from a tentative reactionary state to a concrete vision for the future of the regional municipality." All in all, we were quite impressed with their determination to make the best of a seemingly unmanageable situation.

The settlement of Fort McMurray is nestled in the valley of the Athabasca and surrounded by rolling hills of the northern boreal. A town like any other, it is home to many young families and generations of labour families. There is a great local pride. It's no wonder that so many of the comments on our blog were posted by McMurrians, asking us to actually take a look at their city. It's simply too easy to be misled by mass media intent on uncovering scandal.

– Kalin & Tim

Tonight's Digs: Bruce's basement in Fort McMurray, the day after his fortieth birthday, only hours after we ran into him on the highway and told us he had no place to spend the night.

September 5
Flying Over the Tar Sands

The regional municipality of Wood Buffalo (which encompasses both Fort McMurray and Fort McKay, as well as most of the territory along Highway 881) is not the only community facing the strains of rapid tar sands development. The region of Cold Lake-Bonnyville, an area northeast of Edmonton and southeast of Fort McMurray, also has heavy and increasing rates of tar sands development. In this area, the most common method of tar sands extraction is a process different from the strip mines of the area north of Fort McMurray. In-situ oil sands mining uses deep steam injection to separate the bitumen from the sand and pump it to the surface. It uses intensive seismic testing to locate appropriate well pad sites, drawing gridlines in the landscape: an aerial photo looks more like a giant Sudoku game than boreal forest. Once developed, an area with active in-situ development looks like a giant sprawling suburb, but with no place for anyone to live, including the large mammal species that used to call this area home.

— Lindsay

I was shocked by the dramatic contrast between the open-pit mines swarming with trucks and industrial cities of flashing lights and billowing clouds of smoke, and the serene forest bordering them. But what twisted my stomach into knots were the tailings ponds, the sickly-coloured lake-sized pools, stretching across massive spaces, dwarfing the five-ton trucks that inched along their edges, one built up and looming over the river bank, the river hugging it on three sides.

— Katherine

The flyover was an unforgettable experience, which evoked a great deal of emotion amongst us, but it failed to produce the reaction I myself had expected. After weeks of discussing the issue, the anticipated climax escaped me. From high above, we were able to survey the near full extent of damage in one 45-minute flight — the same amount of time a trip from McMurray to Edmonton or Edmonton to Calgary takes. The destruction stretched far off in every direction, but still it seemed measurable and contained. I remember feeling underwhelmed. I wanted to feel worse than I did, but I'd been desensitized to

the images, overexposed to the issue. I reminded myself that this, the Athabasca, was but one of three regions currently under development and that the plan was for a fivefold expansion in production by 2030. It was only going to get worse.

Then again, an aerial view is a poor measure of scale. I would get a different impression seeing it from the ground.

— Tim

September 6
Cycling Through the Syncrude Plant

The Fort McMurray experience varies, as should be expected. To each his own, right? But what cannot be denied is the gross scar growing on the land a mere 30 kilometres north on Highway 63. The scene is unearthly. A choking stench fills the air. Water levels in the Athabasca River are lower than they've ever been. Residents no longer trust the water, nor the fish that swim in it. Huge plumes of smoke jet out from the towering stacks. Air cannons let out loud blasts to scare the birds to prevent them from landing in the highly toxic tailings ponds that extend well into the distance.

The tailings ponds — which are, in fact, tailings lakes — contain a volume of toxic tailings (leftover water from the bitumen extraction and separation process) that if combined would constitute the third-largest dammed body of "water" in the world.

— Tim

The Suncor site, 2007. The image shows the full range of tar sands industrial activity: upgraders, power plants and industrial facilities in the background, tailings ponds in the middle distance and mining operations in the foreground. On the right, the edge of the boreal forest.

Syncrude's operations in the Alberta tar sands, northern Alberta.

Giant earth-moving equipment is used to mine the bitumen 24/7 at Syncrude, in mines often up to 80 metres deep.

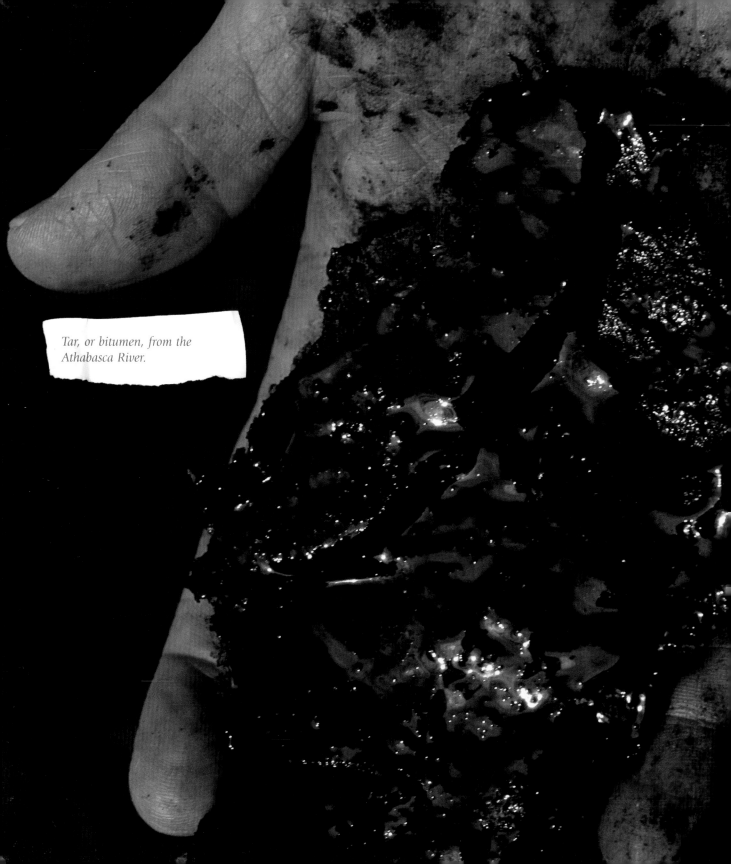

Tar, or bitumen, from the Athabasca River.

At Road's End

Dwarfed by the huge dump trucks

September 7
Fort McKay, A Community Surrounded

The headwind was fierce as we biked past the Syncrude site on the last leg of Highway 63. It felt like pushing up against a brick wall with nowhere left to go; yet while the ride through the tar sands was more or less the end of the line, it was by no means the end of the story. Beyond Syncrude are the communities of Fort McKay and Fort Chipewyan, for whom the tar sands are not the climax to some great adventure, but an everyday fact of life.

Dylan, Kealan and I were the last to arrive at a luncheon that Fort McKay First Nation (FMFN) was hosting to celebrate our journey's end. Most of the as-yet-unmined oil sands fall within Fort McKay First Nation's traditional lands, so this community, our last stop, represented to us not only what was at stake with oil sands development, but the story yet to unfold.

As we approached the beautiful new Dorothy McDonald Community Business Centre in Fort McKay, a couple of men standing at the front doors greeted us.

"You made it," one said, "come on in, there's plenty of food." We parked our bikes and hurried inside to join the gathering and heap our plates with bannock, salad, chicken wings, ribs and potatoes at the buffet. After eating, we met with Chief Jim Boucher in another room in the centre.

Chief Boucher told us that Fort McKay had resisted the oil sands developments all through the 1980s, but when the market for furs bottomed out and the big oil corporations had a number of projects in the region approved, some people in the community started to change their tune — he was one of them.

In those days consultation with Aboriginal communities was completely left out of project planning processes. In an attempt to get their say and secure some benefits for the community, the

Fort McKay First Nation formed the Industry Relations Corporation and partnered in the oils sands development. Now about half the community is employed through oil sands activity. In fact, Fort McKay is said to be the "richest First Nation in Canada." The community appreciates the economic gains that development has brought to the community, but there have been losses too. Many in Fort McKay speak of the tar sands with resignation. With the new Canadian Natural Resources Limited (CNRL) development to the north, the community of Fort McKay is now almost completely surrounded by tar sands development.

"This community has suffered as a result of development," acknowledged the Chief. "We're trying to maintain what we had in terms of hunting and travelling and social structures. . . . We want to maintain our traditional lifestyle, but the question is, from which period? Some people won't remember what it was like to live in the bush in a family unit 24/7 because they didn't experience it."

After our meeting with Chief Boucher, Jean L'Hommecourt, a mother of five and Fort McKay's Industry Relations Corporation's environmental affairs coordinator, offered to show us around. A bunch of us piled in to her car and Jean drove us further up the highway. She pointed out numerous industrial sites, belonging to Suncor, Syncrude and CNRL, among others, as well as the industrial park owned by Fort McKay that serves the tar industry and employs many members of the community. Then she turned onto the narrow road that leads to her mother's trapping cabin. She told us about how her family had lived as her ancestors had done for thousands of years, hunting, trapping and fishing. Jean described these as days when the sense of community was strong: other travelling families would drop by to visit and share the good times.

Then we headed to Jean's brother's cabin, where her family has a sweat lodge. Close to the river and sheltered by the trees, the site for the sweat lodge was beautiful. Walking around I felt the tranquility of that sacred place.

"They're building a camp just down the road," Jean said, as we got back into her car. Someone asked, "Why are all those rocks blocking that road down to the water?"

"The oil companies put them there to control access to the area," Jean replied. "At certain times, they even make us sign in at a checkpoint when we go to my brother's to sweat. That always makes Mom mad."

– Jeh

Tonight's Digs: Gym floor of the band office in Fort McKay.

September 8–9
Trading Pedals for Paddles — The Downstream Effects of Tar Sands Development

After our farewell visit to Fort McKay, our group broke up. We said our goodbyes as each member left for home — some catching rides with friends, others picked up by family, a few cramming in with Jodie, who was returning the car to the rental company in Calgary.

I lingered on in the hopes of getting to Fort Chipewyan, a small First Nations Reserve 200 kilometres north of Fort McMurray. I was determined to get there but by the end of the trip there was only one other taker, Dylan.

You can only reach Fort Chip by boat, plane or ice road, so it wasn't on our official route, but it's an important part of the tar sands story. As oil sands and other industrial developments have increased upriver, Fort Chip has seen a persistent and significant increase in the incidence of

Making our way up the Athabasca

The L'Hommecourt family in Fort McKay

certain diseases among its population. In Fort McMurray, we had met elder Peter Deranger, a long-time Fort Chip resident, who told us about a report by John O'Connor, a physician from Fort Chip, about the disproportionately high incidence of colon, liver, blood and bile-duct cancers in the community. In the general population, the incidence of cancer of the bile duct, cholangiocarcinoma, is 1 in 100,000, but in this community of about 1,200 people, tissue biopsies have confirmed two cases of the rare cancer. One in 600 is a startling enough statistic, but there were another three deaths of people who showed symptoms of cholangiocarcinoma but had not had their diagnoses confirmed by biopsy because of Fort Chip's limited health resources. Peter Deranger told us that this rare cancer is strongly associated with carcinogens such as arsenic and polycyclic aromatic hydrocar-

bons, both of which are discharged as a by-product of extraction activities. Downstream from the immediate vicinity of tar sands development, and therefore enjoying none of its benefits, the

community of Fort Chip has nonetheless paid a heavy cost.

As it turned out, I did not make it to Fort Chip. Instead, Dylan suggested we should stop by Jean L'Hommecourt's for tea. I'm still not sure how she felt about us showing up on her doorstep unannounced that Saturday morning, but she graciously invited us in. We chatted a bit and met her family. When we discovered Jean's family had a boat, Dylan, never the shy one and always a charmer, asked them how they'd feel about doing a little run down the river. The thought of getting on the water was mighty appealing. Jean and her family very generously offered to take us; they would make this their last boat ride of the season.

Dylan and I sat at the front of the small motor boat, the warm sun and piercing wind vying for our attention. Looking behind us, upstream, we could see plumes of smoke billowing from the Suncor plant, but we were headed downstream and from that perspective the Athabasca is a most beautiful river. As our captain navigated through the sandbars dotting the waterway, our hosts commented on the low water levels. They pointed to a new water uptake facility on the river bank, belonging to the massive CNRL project currently under construction.

We spotted a bear in the distance, a good omen. Further along, we stopped and went ashore to explore an abandoned oil sands mine, Alberta's first. Although it is designated a historic site, few people are likely to ever see it, since it is inaccessible by land. I picked up a piece of tar oozing from the river bank. This was the stuff we'd been talking about for the past three weeks, and here it was in my hands.

On the return trip, we eyed a hawk up above and paused to watch several beavers building a dam. The L'Hommecourts told us they no longer trust the water's safety: they don't drink it (they buy bottled water) or eat the fish that swim in it. The industry claims its impact on the river is minimal and that leaks from the tailings ponds virtually never occur. Still, many of the ponds are nestled on the river's edge and in heavy rains there is a risk that toxins will be released into the waterway. A few days later, after I got back to Montreal, I learned that there had been a minor leak on the very day we spent on the water. In such incidents, which tend to be kept quiet, the toxins make their way downstream, passing through places such as Fort McKay and Fort Chip.

After we said goodbye to Jean and her family, Dylan and I spent our last night camping under the stars on a large sandbar across the river from Fort McKay. Over on the far end of town, someone launched some fireworks. Behind us, we could hear the beavers slapping their tails on the surface of the water. We made a fire with driftwood and prepared a feast with the food remaining from the trip. The next day, in a canoe lent to us by the L'Hommecourts, we paddled upstream and onto the Red River, a small tributary of the Athabasca. We were now as far north as we would go. The early fall scenery was spectacular, and as we drifted along with the current, I sat in the bow in complete awe of my surroundings, impressed by nature's resilience in the face of the violence currently being inflicted upon it. I felt a calm, a tranquility. For the first time in weeks, I stopped looking for "the story" and allowed myself to truly be in the moment.

– Tim

Afterword: Standing at the Crossroads

We're all back home now, or off to other adventures. Almost a year later, we're still digesting our experience.

In Canada, we truly sit in the belly of the beast. With $146 billion of investment anticipated in the next decade, Alberta's tar sands are the biggest mega-industrial project in the world. The expected growth in greenhouse gas emissions from the oil sands alone will prevent Canada from reaching its targets for reductions. And thousands of boreal ecosystem kilometres are being rolled back by the biggest trucks and equipment humanity has ever seen.

Yet our elected leaders have failed to respond adequately. Since our ride, the Conservative government of Ed Stelmach was re-elected with a strong majority; the controversial Bill 46 (restructuring the Energy and Utilities Board) was passed into law; resource royalty rates were raised and then re-adjusted in response to corporate pressure; wildcat strikes were staged across Alberta as workers protested laws they claimed limited their right to strike; a precedent set by a Queen's Bench decision entitles landowners to water testing before any new oil and gas developments begin; environmentalists claimed a huge legal victory with the halting of the Kearl Oil Sand Project, only to have the court's decision overturned as the federal government fast-tracked a missing water permit; and hundreds of birds died after landing on Syncrude's tailing ponds because the noise-making cannons had not been deployed in time for the migration. Victories to date have been few but the silent majority is finding its voice and a movement is building.

Indigenous groups including the Mikisew Cree and Athabasca Chipewyan First Nations in Fort Chipewyan have called for a moratorium on oil sands development. Development displaces indigenous peoples from their traditional lands and because their communities and economies depend on the land they are especially affected by the environmental degradation associated with tar sands development. A grassroots organization called the Indigenous Environmental Network (IEN) is addressing such environmental and economic justice issues. The IEN stresses that harm to indigenous communities caused by the development of non-renewable resources begins with exploration and expands with the building of industrial infrastructure. Aboriginal treaty rights supersede the rights of the Province of Alberta and of corporations operating within Canadian borders, so by approving new projects while First Nations call for a moratorium, the federal and

provincial governments are breaking the law and sacrificing people's health and well-being for economic gain. When governments fail to follow their own laws, all Canadians are at risk. If there is one thing that our experience biking to the tar sands taught us, it is that a relationship of solidarity with indigenous peoples is crucial for a just resolution to the problems associated with the tar sands development.

For three weeks last year, we listened to people's stories and lent our support however we could, only to pack up and ride off the morning after. But for the people we met, the tar sands don't represent some sexy bike trip, but a day-to-day struggle. As observers, it is our responsibility to communicate their stories effectively. The power and privilege we hold as activists fuelled an emotional debate among our group as we debriefed the experience over email later. What had we accomplished other than a slide show full of interesting pictures and stories to tell from our summer spent on a bike?

In the end, as allies of the people living with the effects of the tar sands we must not only support the battles for environmental justice going on in communities such as Marie Lake, Fort Saskatchewan, Fort McKay, Fort Chip and in backyards across this country, but also challenge the oppressive and systemically racist structures that pervade our own lives.

The issues surrounding the tar sands speak of our struggle to put food on the table, our inability to break free from consumerism, our delusions of wealth and happiness, our fear of change. No one really cares for oil — only for the things it produces and the services it provides. We are at a crossroads, with a choice between the hard path, characterized by unchecked growth, indiscriminate consumption and dirty energy, and the soft path, characterized by voluntary simplicity, greater efficiency and renewable energy.

Facts About the Tar Sands

- Oil sands are contained in three major areas of northeastern Alberta beneath about 140,000 square kilometres of muskeg and boreal forest.
- Most oil sands deposits contain roughly 10 to 12 percent bitumen, 80 to 85 percent sand and clay, 4 to 6 percent water, and small amounts of heavy metals and other contaminants.
- Bitumen is a thick, sticky, molasses-like substance, suitable before processing only for paving roads, and containing up to 5 percent sulphur. Compared to crude oil, bitumen contains too much carbon and too little hydrogen, which is corrected in upgrading.
- In 2006, bitumen represented only 31 percent of Alberta oil and gas production on an equivalent energy basis. However, bitumen resources account for 95 percent of remaining established reserves.
- One barrel of synthetic crude oil from the tar sands requires 2 to 5 barrels of water to extract the bitumen.
- Every 1.0 cubic metre of oil sand mined creates 1.4 cubic metres of material for disposal, because the particles do not re-consolidate.
- The extraction of one barrel of bitumen produces six barrels of tailings.
- In one day the oils sands industry produces 1.8 billion litres of tailings.
- On average there is one leak in a tailings pond dam each year.
- Tailings consist of a slurry of water, clay, sand, residual bitumen and chemicals, including naphthenic acids, which are acutely toxic to organisms. Naphthenic acids are present in concentrations between 40 to 120 milligrams per litre.
- On one day in April 2008, 500 ducks died after landing on one Syncrude tailings pond, at least in part because they became contaminated with oil.
- Up to 1.2 million barrels of oil are produced from the tar sands per day.
- The energy equivalent of one barrel of oil is required to produce three barrels of oil from the tar sands.
- Production of one barrel of crude oil from the tar sands results in the release of approximately 100 kilograms of greenhouse gases, compared to 23 kilograms for one barrel of conventional crude.

- Less than 20 percent of the oil sands can be mined; the rest of the recoverable deposits are too deep, and so are recovered using in-situ methods.

- In-situ methods of recovery extract the bitumen from the sand while it is still in place through the addition of heat, by injecting steam into the deposit to allow the bitumen to flow out.

- Approximately 500 square kilometres of land have been disturbed by oil sands surface mining activity; the lease holders are obligated to reclaim this land, that is, to restore it to the equivalent of its original state.

- In 1993 Syncrude moved a herd of wood bison to reclaimed land in a grazing research project in co-operation with the Fort McKay First Nation. Today, the herd numbers approximately 300 head.

- Syncrude operations have disturbed 18,335 hectares of land. Syncrude spends more than $14 million each year to reclaim land disturbed by its surface mine, and claims it has reclaimed 4,500 hectares.

- Suncor's activities have disturbed approximately 13,000 hectares. To date, it claims to have reclaimed approximately 1,000 hectares of land (roughly 7 percent of the land disturbed), which now support a diversity of wildlife and native plants.

- In 41 years of tar sands mining, the Alberta government has issued one certificate of reclamation (signifying that reclamation is complete to Alberta's standards) — to Syncrude, for Gateway Hill, a 104-hectare site that had been used as a soil dump, 35 kilometres north of Fort McMurray (approximately 0.2 percent of land disturbed by the industry).

- Albertans receive 32 percent of the net revenue from oil sands development through taxes and royalties; companies take 53 percent; 15 percent goes to the federal government through income taxes.

- Over the past three decades, the Alberta government has invested close to $1 billion in oil sands research programs jointly with industry.

- The oil sands account for more than two-thirds of investment in Alberta. One hundred billion dollars' worth of oil sands investment generates approximately one trillion dollars' worth of economic activity.

- Alberta produces about 81 percent of Canada's natural gas and 68 percent of its crude oil.

- 145,000 Albertans are employed in the mining and oil and gas extraction industry. Thousands more work in the services sector that supports energy exploration and production.

- According to Statistics Canada, Alberta's population increased by 103,400 people in 2006/2007. At a rate of 30.2 per 1,000, Alberta's population is increasing more than three times as fast as Canada's.

- Alberta has had the highest rate of economic growth, 4.3 percent, in Canada over the past ten years. In 2006 Alberta's economy grew by 6.6 percent.

- According to an April 2007 poll, 71 percent of Albertans believe that the Alberta government should suspend new oil sands project approvals until environmental and infrastructure issues have been resolved.

- The United Nations Environment Program has identified Alberta's tar sands mines as one of 100 key global "hotspots" of environmental degradation.

- Often overlooked, Saskatchewan is Canada's second-largest oil producer. The Alberta oil sands boom is spreading eastward to the oil sands of the Clearwater River Valley.

Sources: Canadian Encyclopedia, Canadian Water Network, Environmental Defence, Environmental Services Association of Alberta, Kealan Gell, Globe and Mail, Government of Alberta, Nexen Inc., Oil Sands Discovery Centre, Oilsands Infomine, Oil Sands Review, Pembina Institute, Polaris Institute, Sierra Club, Suzuki Foundation, Suncor, Syncrude, Tyee.

The Contributors

Jeh Custer

Jeh recently moved to Edmonton to work with Sierra Club Canada on the Tar Sands Timeout and Mackenzie Wild campaigns (www.tarsandstimeout.ca and www.mackenziewild.ca). He is helping organize a second youth-led tar sands bike trip called Return to the Tar Sands.

Gregory John Ellis

Greg is an English major at Cal Poly in San Luis Obispo, California, and spends his time planting trees, organizing community dinners, romping around in compost heaps, playing music, reading poetry and distracting himself and others from the urge to drive around and buy stuff.

Aftab Erfan

Originally from Iran, Aftab grew up in West Vancouver, went to school in Vancouver and Montreal, spent two years living and working in Halifax, and is currently employed as an urban planner for the district of North Vancouver. She has been involved in the youth environmental movement for the past eight years, as a student with the Sierra Youth Coalition, as a director of the Youth Environmental Network and as a founding member of the Canadian Youth Climate Coalition (http://www.ourclimate.ca).

Jacqueline Gamble

Jackie has worked on a mouldy circus theatre ship in Holland, experienced first hand the electrifying insanity that is World Cup, learned to drive a mule using Berber commands in Morocco's Atlas mountains and worked for an NGO in southern Sudan. During the bike trip, she worked for ondah.com, a new medium for spontaneous personal expression through mobile photography.

Kealan Gell

Kealan grew up in Sooke, British Columbia, studied engineering in Quebec and Holland, and fell in love with Katherine during this bike trip. He uses music and labour to release his ideas about economic de-growth and closed loop agriculture.

Jodie Martinson

Jodie Martinson, a native Calgarian, is a documentary filmmaker who pedalled to the tar sands with a video camera strapped to her saddlebags. She is pursuing a master's of journalism degree at UBC. Her feature-length documentary about the trip is available by contacting her at saddlebag.productions@gmail.com.

Tim Murphy

Tim developed an interest for energy issues while cycling across Canada with the Climate Change Caravan. He is a former coordinator of the Sierra Youth Coalition's Community Youth Action Project, and past member of its Executive Committee. Born and raised in Moncton, New Brunswick, Tim lives in Montreal, where he works as the Sustainability Coordinator for an amazing organization called Santropol Roulant (http://www.santropolroulant.org/), whose mission is the delivery of nutritious meals to individuals living with a loss of autonomy. His work centres around Santropol Roulant's Sustainable Urban Food Cycle and Rooftop Garden Project.

Dylan Sparks

From his experiences in the wild spaces of British Columbia, Dylan has developed an unabated passion for the Earth, committing his actions and spirit to the preservation of our mutual home. A gifted athlete, Dylan is drawn to the "active" component of the social justice and environmental movement. This unique combination is part of what made the tar sands project so appealing to him. His inspiration from the work of the Sierra Youth Coalition and his desire to gain experience in the non-profit sector led Dylan to this incredible project.

Kalin Stacey

Kalin Stacey is 22 years old and lives in Toronto. He is currently an undergraduate student at the University of Toronto, specializing in Linguistics and Equity Studies. He is infinitely grateful for the learning he gained on the bike trip, and hopes that these tales will educate and inspire others as well.

Lindsay Telfer

Prior to starting her new position as Prairie Chapter Director, Lindsay was working alongside the Sierra Club of Canada's National Office as a grassroots organizer. In this capacity, she has been coordinating and conducting trainings on fundraising, strategic planning and campaign development across Canada. She was the National Director of the Sierra Youth Coalition from 2002 to 2004 and is also co-founder of Energy Action, a cross border youth coalition for clean and just energy. Lindsay has a master's degree in community-based food systems from the Faculty of Environmental Studies at York University and has a successful history in campaign and finance development as well as strategic planning.

Katherine Trajan

Brought up in beautiful British Columbia. Imagined that the whole world was pristine and stunning. Went to school. Lived in cities. Travelled. Read. Realized that such rare beauty must be protected. Trip to the tar sands a lingering memory.

Lori Theresa Waller

Lori is a writer who lives in Ottawa, where she watches geese, grows vegetables, plays drums and dances. In former lives, Lori studied sociology, worked as a vegetarian cook and biked from Saskatchewan to Nova Scotia with the Otesha Project.

Fellow Travellers:
Jonathan Braz, Peterborough
Joanna Bruijns, Hamilton
Joanna Dafoe, Toronto
Maya Dion, Montreal
Emma Hamilton, Ottawa
Shawn Khan, Edmonton
Robert Larson, Vancouver
Danny Spitzberg, Boston